A PARENT'S GUIDE TO STEM

CONTENTS
A PARENT'S GUIDE TO STEM

STEVE DEBENPORT – GETTY IMAGES

CHAPTER FOUR

Map a Path Forward

COVER PHOTOGRAPH: Steve Debenport –
Getty Images

The

S

SCIENCE

T

TECHNOLOGY

Adva

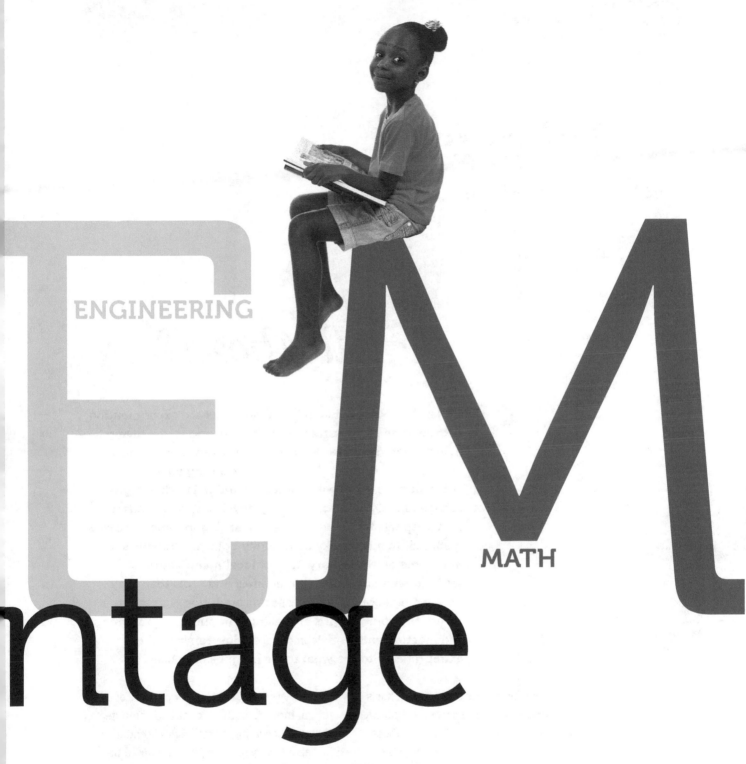

ENGINEERING

MATH

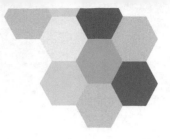

What Is STEM *All About?*

 In New York City public schools, students are experimenting with solar energy in urban greenhouses and scuba diving in local waters to experience environmental science up close. Fifth-graders in Georgia are proposing ways to redesign a local zoo with plants and animals, while high schoolers in Ohio are shadowing physicians on rounds at area hospitals to see what they're learning in biology classes firsthand. At a Kentucky community college, students spend more time on the factory floor of local manufacturers working with cutting-edge machines and tools than they do learning about the processes through books or lectures. Across the country, girls are taking computer programming and meeting with female software engineers and tech entrepreneurs to see what these hot jobs look like.

It's no secret why there is so much emphasis at schools nationwide on science, technology, engineering and math, or "STEM," as those fields are often referred to collectively: Simply put, that's where the jobs are. From coast to coast, employers big and small are actively seeking workers with specific STEM skills. Companies can't find qualified employees to fill the many job openings that exist today, and they are worried that there won't be enough

tech-savvy employees for the workforce of the future. Indeed, math, science and computer skills are becoming increasingly essential for just about every industry, from manufacturing to health care to finance to education. Government and business leaders are championing the innovative power of STEM as vital to the nation's economy and looking for ways to equip students and professionals from all backgrounds with these abilities.

Job opportunities

Why is the United States facing such a critical dearth of employees with STEM skills? In short, far too few youngsters have been choosing to take science and math classes and to stick with the subjects in college. Some students show promise in these disciplines but often lose interest; others are not being exposed to role models or mentors. Nor are students getting enough opportunities to gain valuable hands-on experience in these fields. As a result, educators, employers, government officials and other advocates are making a strong push to get children, particularly those from underrepresented groups and underserved communities, involved in science and math.

There's good news for your child. In the years ahead, jobs in the STEM fields are projected to grow about twice as fast as those in other industries. What's more, many of these careers are among the highest-paying: College STEM graduates typically make about $500,000 more over their lifetimes than their peers who studied other subjects. And many STEM jobs don't even require a four-year degree, but instead are being filled by those with industry certifications or degrees from community colleges that often partner with businesses to make sure workers learn exactly the right skills needed for the workforce.

But getting a good job isn't the only reason you should help your child develop a solid grounding in STEM. In an age of smartphones, the Internet, remote health care and computers, your child needs to master basic STEM skills to thrive in today's world. A little bit of engineering and computer science can go a long way no matter one's future career. The key is to get kids excited about science at an early age –

even as young as preschool – and find ways to help them keep learning it. Dozens of schools across the country are transforming their curricula to better tackle STEM and help ensure that students are college- and career-ready. Numerous after-school and summer programs are cropping up that give students of all ages and backgrounds extra access to tutoring, mentoring and hands-on science activities – from building robots to designing mobile apps and websites to performing basic experiments in biology and geology.

How you can help

Fortunately, you can be one of your child's best advocates for STEM, no matter what your own educational background is. That's why U.S. News & World Report has created "A Parent's Guide to STEM." On the following pages, you'll find tips on how to introduce youngsters to science and math by, say, taking a nature field trip, cooking or reading a biography of female scientists; advice on picking the right high school and classes for your child; and words of encouragement from young professionals and STEM role models. You'll also find some leads to help you and your child launch a college search, an idea of what students can expect from the new SAT, a sampling of STEM scholarships, a list of hot STEM college majors and advice on how to land an internship.

Perhaps the best strategy of all: Find a way to make STEM fun and help your child understand how it works in the real world. He or she could soon be on the way to becoming a biochemist, video game designer, physicist, electrical engineer, automotive technician, cybersecurity analyst – you name it. The opportunities are just about endless. ●

From software developer to biomedical engineer to environmental scientist, STEM jobs are exciting and rewarding. Here's a sampling of some of the in-demand STEM careers that you and your child might want to explore:

> *Among the 25 top-paying college majors, 23 are in STEM fields and all lead to median salaries of at least $76,000 a year.*

SCIENCE

- Archaeologist
- Astronomer
- Audiologist
- Biochemist
- Dietitian and Nutritionist
- Environmental Scientist

- Epidemiologist
- Exercise Physiologist
- Food Scientist
- Genetic Counselor
- Geoscientist
- Marine Biologist

- Meteorologist
- Park Ranger
- Physician Assistant
- Surgeon
- Veterinary Technician

> *In STEM fields, there are approximately two job openings for every one unemployed person.*

TECHNOLOGY

- Animator
- Automotive Technician/ Mechanic
- Cloud Architect
- Computer Programmer

- Cybersecurity Analyst
- Database Administrator
- Game Designer
- Health Information Technician

ENGINEERING

- Aerospace Engineer
- Architect
- Biomedical Engineer
- Civil Engineer
- Construction Manager
- Environmental Engineer
- Highway Designer
- Mapping Technician
- Nuclear Engineer
- Petroleum Engineer
- Robotics Technician
- Transportation Planner
- Urban Planner
- Water/Wastewater Treatment System Operator
- Wind Turbine Technician

Petroleum engineers earn median salaries of $130,000 annually, while software developers make about $93,000.

About half of all STEM jobs don't require a four-year college degree.

MATH

- Accountant
- Actuary
- Biostatistician
- Budget Analyst
- Cryptologist
- Data Scientist
- Economist
- Electronic Math Modeler
- Financial Planner
- Mathematician
- Operations Research Analyst
- Statistician
- Stockbroker
- Surveyor
- Survey Researcher

Women make up less than a third of all STEM jobholders.

- IT Manager
- Medical Technician/ Technologist
- Mobile Developer
- Nuclear Technician
- Social Media Manager
- Software Developer
- Sonographer
- Web Developer

Sources of statistics (left to right): Change the Equation; Georgetown University Center on Education and the Workforce; Brookings Institution; Bureau of Labor Statistics; National Science Foundation

Jobs That Don't Require a Degree

What do MRI technologists, electricians and computer support specialists have in common? All of these STEM jobs are expected to grow faster than average in the years ahead, and each occupation pays close to $50,000 or more annually, on average. Perhaps more surprising: None of these jobs requires a four-year college degree. Many terrific STEM opportunities, in fact, are in the so-called middle-skill occupations – jobs that require more training than a high school diploma, but less than a bachelor's degree. Such positions represent more than half of all jobs that exist today, as well as about half of the careers that will be in demand by 2022, according to the National Skills Coalition.

Why are these skills in demand?

New technology. And it's transforming a number of fields, from health care and telecommunications to IT and manufacturing. Today's medical lab technicians, for instance, who analyze blood and other samples, must be familiar with cutting-edge medical equipment and electronic health records in hospitals, clinics and other settings. Machinists and engineering technicians must also be familiar with sophisticated computer software and robots that are a common part of their day-to-day work. Dental hygienists, carpenters, automotive technicians, heating and air conditioning mechanics and many other workers in STEM-related jobs also require specialized training in new tools and skills.

Many companies want to ensure that they can find people with the specific computer and technical abilities they require. They often partner with community and technical colleges to provide the exact training their workers will need – and even help the trainees move directly into jobs (see also "Weighing the College Option," page 58). Nationwide, thousands of firms are preparing technicians and middle-skill employees by enabling them to learn as they work through apprenticeships and other programs.

To get a sense of these jobs in your area, you and your child can start by looking into what kinds of degree and certificate programs are offered at local community colleges. State and local college-industry partnerships can also provide some leads about in-demand jobs. The North Carolina Advanced Manufacturing Alliance, for example, which includes 10 community colleges and dozens of employers, is helping train new professionals in industrial systems technology and computer-integrated machining so they'll be ready to hit the ground running in factories and plants across the state. ●

5 STEM *Myths*

That You Shouldn't Believe

Popular wisdom is often wrong when it comes to STEM. Here are just a few of the common myths that don't hold up:

. . . .

1 STEM is not a field that welcomes minorities and girls or women

To the contrary, employers hungry for high-tech talent are determined to attract minority students and women into STEM fields. They've launched all sorts of efforts, including internships for high school and college students and partnerships with community colleges that offer the chance to earn an associate degree along with on-the-job training. And while studying STEM can seem like a complex and lonely proposition, there are plenty of resources and support systems, such as student clubs and college learning communities, where, say, female computer science students or aspiring engineering majors live and take classes together and build relationships to help them thrive.

2 Girls don't like math and science

Not true. In fact, girls tend to get better grades in math and science than boys do and generally take the same number of credits in these disciplines. Often, they lose interest or get discouraged when parent or teacher attitudes – or their friends – steer them in other directions. But research has shown that, if girls receive encouragement and support, they often have both the deep interest and ability to do really well.

4 Science and math are boring and not relevant to real life

In fact, everyone has to draw on his or her science and math skills in myriad ways each day. Parents and teachers can emphasize this reality as they apply measuring and chemistry in cooking, compare unit prices at the grocery store, explain how a car engine works and so on.

3 Math and science are for brainiacs only

That's the old "weed out" mentality – a long-held assumption by some science and math teachers and professors that most students in lower-level classes will not make the grade and will take another path. But that attitude is changing fast, as studies have shown that a passion for STEM subjects is much more key to success than sheer brainpower. Old teaching methods like boring lectures and memorization – which rarely inspired any enthusiasm – are rapidly being replaced by more exciting and success-oriented approaches in the classroom. Project-based learning, for instance, requires hands-on tackling of real-world challenges; students might develop new renewable energy sources, say, or use algebra and geometry to design furniture. That sort of engagement, along with the right grounding in key skills like math, provides the pathway to success.

5 STEM jobs are isolated and lonely

Not so. There are hundreds of STEM professions, and many require people to team up to solve problems – as environmental scientists working to clean the water and air, say, or engineers cooperating to design new space vehicles, or public health professionals working with communities to wipe out disease. These workers can make a difference in the lives of millions.

Aprille Ericsson
Aerospace engineer, technologist and instrument project manager, NASA

What sparked your interest in STEM?

My mom was a teacher, so I sort of walk behind her shoes. A number of women in the family were teachers too, and the men were engineers. My grandfathers included me [in activities] like getting up on a scaffold and helping to scrape down the house, or fixing a doorknob lock. Coming up with solutions and using your hands to build and make things, I think, were all good assets for me to become an engineer.

[Going to school in New York City,] I did lots of science classes. I won second place in the science fair in eighth grade. My experiment was a homemade glass milk barometer that I made with my grandfather. I've been at NASA's Goddard Space Flight Center 20-plus years; half of my career here, I was an instrument manager. The fact that I started building my first instrument in junior high school and that's what I ended up doing is pretty cool.

I also believe very strongly in being well rounded. For me, sports were so important. Working on a team, being a leader, having versatility, and then learning to be a good teammate are all skills that I think help me in my current career. People forget that engineers aren't just in the room by themselves working on a design. We work on huge missions with multiple people in multiple countries, which brings in another aspect of being well rounded: having another language or having learned about other cultures.

What is one of the biggest challenges you faced – and overcame – in STEM?

I went to the Massachusetts Institute of Technology without having calculus. I stumbled through math my first year, and then it caught up with me my second year when I had to take differential equations. You really need to know them to be able to do applied math problems. I failed that class twice. So I went to The City

College of New York for a semester and took classes. I did really well, and then I got a 98 on the final for differential equations. That was a big boost to my confidence. It also goes back to what we tell freshmen going into engineering: how important it is to master the tools and required coursework before moving on to the upper-level coursework.

How can parents help?

Kids can go to museums and do hands-on activities. I went to the Brooklyn Public Library, the botanic garden, the Brooklyn Museum. There are lots of professional organizations, like the National Society of Black Engineers or the American Institute of Aeronautics and Astronautics, that have children's activities. There's so much free stuff out there. You just have to get out of the house. Or when your child has a science question, get on the Web when you're not sure of the answer.

Let your children cook. That's being a scientist. I learned how to measure and bake. I did sewing in junior high school, and I learned how to put things together, how to make a pattern, which is very much what an architect is doing when laying out a building or designing a bridge. I did Girl Scouts. All of those experiences as a child helped to build up my skills, so that I could pull off these bigger projects that I do at NASA. ●

MyStory →

Chris FitzSimmons
Instructor, Birch Aquarium at Scripps Institution of Oceanography, UC–San Diego

I chose my career when I was 3 or 4 years old. I was taken to an aquarium and was fascinated watching the staff care for the fish. My mother and my grandmother saw my interest and encouraged it. They got me books on marine life and took me to the Philadelphia Zoo and the beach. I watched television shows on ocean life and in first grade wrote that I hoped to be a marine scientist at a place like Scripps Institution of Oceanography.

My mother found a great vocational agricultural high school in Philadelphia for me with aquaculture, environmental science and a working farm. We studied and cared for fish, horses and other animals as well as the farm. Through a school internship, I became a 14-year-old exhibit guide at the zoo. Wanting to pursue marine science, I attended Stockton College (now Stockton University) in New Jersey and got a lot of hands-on experience. For one tropical marine biology course, our class went to the Florida Keys, exploring the Everglades, scuba diving, snorkeling, and seeing the incredible diverse life there. My instructors were amazing. I saw how they inspired people to learn more about the ocean, which is how I drifted into education. Today, I live my childhood dream at Scripps' Birch Aquarium, educating adults and kids – many who have never before seen the beach – on the oceans and their wonders, and the need to preserve them. ""

Stephanie Reeves
Engineering Advisor for Facilities Engineering, Chevron

After the 9/11 attacks, I decided to apply to the U.S. Military Academy. At home, I had always been at the top of my class, but at West Point I was surrounded by brilliant classmates. My microbiology teacher once said to me: "Stephanie, you may not be the naturally smartest person, but you're the hardest-working person I've ever met in my life, and that will take you farther." I didn't especially appreciate the comment, but in some ways my willingness to work hard has helped me overcome a lot of challenges.

After graduating from the Point with an engineering degree, I was enjoying a fast-rising military career when it was cut short by serious complications from knee surgery. As a captain, I had gained great leadership experience, but no longer had the chance to immerse myself in chemistry or systems engineering, so returning to those fields was a challenge. A military careers conference led me to an oil company, which hired me as a production operations engineer. I had a lot of operations experience, preparing soldiers for deployment, but I had to give myself a crash course in petroleum. I bought textbooks and studied. I found mentors and people to go to for advice. I took a petro-skills course offered by the industry in Houston and got up to speed.

Now at Chevron, I combine my leadership and technical skills to manage cross-functional teams – IT, operations, engineering – to come together to solve problems. It's a satisfying mission. ""

Reshma Thakore
Senior Scientist, Alcoa

I never imagined myself being a scientist. When I was in school, because I was outgoing, I thought I would go into a field like communications. But I always had the science bug – a childish sense of curiosity. When I was a girl, I had this doll, Li'l Miss Makeup. If you put a sponge with cold water on her face, makeup appears. I had the idea of burying my baby doll in snow and realized I could make her have makeup all day. Despite this, I never connected with science in my classes. The "textbook" version of STEM didn't engage me, so I didn't get particularly good grades in science through high school. But when I was at the University of Pittsburgh, I took a chemistry class to fulfill a science requirement. I completed my first exam and got 96 percent. My professor told me, "I think you have a knack for chemistry and science" and suggested I check out the research labs to see if there was a project I would enjoy working on.

I ended up doing bioanalytical research into neuropeptides. The hands-on work really engaged me. I was hooked. Eventually, my research experience helped me get a job at Alcoa, where I focus on different research and development projects. For example, I work on our EcoClean titanium dioxide coating that is placed on buildings and actually breaks down or "eats" smog. Basically, I think for students to really know if they might enjoy science, they should get hands-on experience wherever they can. It made the difference for me. "

Jose Romero-Mariona
Lead Research Scientist for Cybersecurity, Space and Naval Warfare Systems Command Center Pacific

At 16, I came to the U.S. from El Salvador and moved from English immersion classes to regular and AP courses. Math always intrigued me – how you could take different paths to the same answer. I felt the same way about computers, which I often used in the library since my family couldn't afford one. I discovered software is like Play-Doh. You can add things to it, roll it into different shapes and put it away when you're done. And it also doesn't require much money or equipment. I excelled in school, but my guidance counselors discouraged me from applying to college because of my background, so I talked to friends who were being groomed for college. I learned about the courses I needed and about financial aid. I got into the University of California–Irvine where I earned my bachelor's, master's and doctoral degrees. The school gave me great support.

I finished my Ph.D. on a fellowship that enabled me to become a government scientist. Today, I lead research and development in cybersecurity, finding new ways to keep bad guys out of computer networks running critical infrastructures – like water purification, electrical grids, and oil and gas refineries. I've learned in life never to simply accept someone else's "no." It just means you may need to try again, try harder or take another route. "

Get
Insp

ired!

A Journey of Discovery

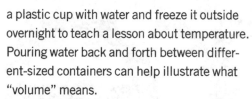

If you want to inspire an appreciation of science and math in your kids, you don't have to go farther than your home or a city park. A simple walk outside can become a journey of discovery, whether it involves grabbing a dandelion puff to blow or examining a caterpillar and discussing how it will soon become a butterfly.

By feeding your child's natural curiosity, you can foster a love of learning and exploration that can lay the groundwork for studying and excelling in science and math at school. Research has shown that kids' fast-growing brains allow them to take in huge deposits of information. They are eager to learn, and supporting that hunger is key to setting them up for future success. Here are some ideas for converting commonplace items and experiences in your own home and neighborhood into learning adventures.

Turn your home into a laboratory

Almost any common utensils can be used for ready-made science experiments or math lessons. For example, in the winter, you can fill a plastic cup with water and freeze it outside overnight to teach a lesson about temperature. Pouring water back and forth between different-sized containers can help illustrate what "volume" means.

Your local library boasts a number of books that can help you conduct science experiments at home, such as "The Kid's Book of Simple Everyday Science" ("Curl Up With a Great Read," Page 25). You can also tap into free interactive resources such as Code.org, a site that enables kids as young as 4 years old to learn the basics of computer science by doing coding exercises. For example, one activity invites kids to use the character Elsa from the Disney film "Frozen" to create a snowflake.

Cooking is another great way to engage kids, suggests Sophia Kraus, interim director of the Center for Child Development at

the University of Akron in Ohio. You can ask, "What can we put into this? How can we change the taste? Why does the egg look different once it's in the frying pan?" Cooking also requires lots of measurements, which can help build math skills.

Explore the outdoors

If you and your child take a ramble together in your backyard, neighborhood or local park, it can spark a good conversation about nature, from fallen leaves to bugs. Ask your youngster to try identifying the different birds he hears or sees. Pick wildflowers and save them to identify later using a book or the Web. Preschoolers love to peer at insects and worms that can easily be found by overturning rocks or

soil. Bees can be observed pollinating flowers.

For school-age kids, the outdoors gets a lot more interesting with an inexpensive magnifying glass to get an up-close look at tiny critters. Have your son or daughter keep a journal to record what he or she finds and use it as a starting point to learn more about insect habitats and behaviors.

On clear evenings, go outside and have your kids notice how the moon and stars have changed their positions. You can even download free apps on your phone to help identify moon phases, stars and constellations.

Learn together

Regardless of the activity, parents should act interested themselves, not only by answering their children's questions but also by asking their own. Don't be afraid to say "I don't know." Looking for answers together can be a bonding experience. Often you can type direct questions into Google, like "Do butterflies sleep?" and find an answer. (They apparently don't sleep like humans do, but they do rest.)

A reference librarian can help you find books to dive into subjects more deeply. There are excellent specials and series on television and the Web as well that cover everything from nature to space exploration to engineering marvels ("Standout Science Shows," Page 28).

As you spend time with your child, avoid reinforcing stereotypes about what boys and girls should do. Refrain, for example, from shuddering at the bugs or spiders kids find. Encourage each youngster to try the activities and be supportive when they do.

Studies have shown that engineering stu-

Many sponsor science programming designed specifically for kids. The North Carolina Museum of Natural Sciences in Raleigh, to name one example, offers a range of free family activities, from game night to question-and-answer sessions on dinosaurs.

You can also check with your local schools, libraries and community centers for after-school or other programs like Family Engineering Nights, which are supported by the National Science Foundation. These entertaining, hands-on workshops introduce children and their parents to engineering principles. Families might find themselves, for example, designing "helmets" for raw eggs to be dropped from 10 feet without breaking or building towers out of pipe cleaners. (Many locales also host Family Science and Family Math Nights.)

Local colleges and universities may offer more advanced opportunities and field experiences for school-age children. The University of Minnesota College of Science and Engineering, for one, offers several STEM-focused programs, from a free, three-day technology day camp for sixth- to eighth-graders that focuses on robotics, to Kids University, a weeklong math and science camp covering topics ranging from rocketry to electrochemistry. There may be fees, but scholarships are often available.

Most important, parents should try to stay involved in activities by listening and asking questions, while letting the kids take the lead. Research shows that when parents are engaged in their kids' learning, children do better in school. ●

dents exposed to building or mechanical tasks as young children tend to perform much better, so you want to give both boys and girls that opportunity. Children will inevitably pick up on gender stereotypes through other kids, adults and TV. You can counter those messages by making sure your sons and daughters are exposed to all sorts of play.

If you can only afford one building toy, experts suggest Lego Classic building bricks. Playing with Legos teaches youngsters to think in three dimensions, helps them develop problem-solving and communication skills, and boosts motor development.

Take advantage of community resources

As children reach school age, keep the momentum going by visiting museums, planetariums and science centers ("Take a Field Trip," Page 21), some of which offer free admission on certain days or allow you to pay what you can afford.

Take a Field Trip

Kids learn best through hands-on experiences that encourage them to figure things out for themselves. So especially with STEM subjects, teachers and schools plan activities that allow for discovery – and so can parents. The following facilities, which range from fossil sites to space centers to factories, are all within field-trip distance of major cities. You can probably find unique opportunies close to home, too. Ticket prices for the facilities below range from free to $35. And remember: Many museums have free days or let you pay what you can afford rather than the posted admissions fees. So plan an adventure with your children and let them explore!

ATLANTA

- **Chattahoochee River National Recreation Area,** Sandy Springs ($). Bike or kayak down the 48 miles of river in this national park, home to blue herons, otters, owls and bats, among other wildlife. And sit in on a ranger-led program on forest and river ecosystems.
- **Centers for Disease Control and Prevention Museum** (free). The CDC's permanent exhibits include the electron microscope used to study bird flu, West Nile Virus and AIDS, and the needle-free injector integral to reducing smallpox outbreaks.
- **Fernbank Museum of Natural History** ($).

» **Hands-on exploration at Boston's Museum of Science**

BOSTON

- **Nash Dinosaur Track Site and Rock Shop,** South Hadley ($). At the first valley in North America where dinosaur tracks were discovered, walk the same paths as dilophosauruses and plateosauruses and spot their footprints.
- **Blue Hill Observatory and Science Center,** Milton ($). Learn about atmospheric science in a building dedicated to its study since 1885. For an extra fee, get a lesson in weather forecasting or kite-making.
- **The Hall at Patriot Place,** Foxborough ($). Investigate the math and science behind sports. Activities include measuring force and angle to determine how far a ball will travel and designing the most effective helmet.
- **Museum of Science** ($).

CHICAGO

- **Willowbrook Wildlife Center,** Glen Ellyn (free). A rehab facility for injured and orphaned animals, with exhibits on Illinois wildlife and several nature trails.
- **Kent Fuller Air Station Prairie,** Glenview (free). This 32-acre

prairie re-creates a time when much of North America was covered with tall prairie grass. It showcases crayfish, red-tailed hawks and 160 plant species.

- **Museum of Science and Industry** ($). Explore the connection among different sciences through an interactive coal-mining exhibit, a simulation of future technologies, and more.

DALLAS

- **Dallas Arboretum and Botanical Garden** ($). The 66-acre facility has 19 gardens and programs for kids such as "Wetland Wonders" and "Animal Homes."
- **In-Sync Exotics Wildlife Rescue and Educational Center,** Wylie ($). Dedicated to helping injured and abandoned exotic cats, the refuge houses cougars, tigers, lions, leopards, cheetahs and more.
- **Mineral Wells Fossil Park,** Mineral Wells (free). Dig up fossils dating back 300 million years and take them home. Sea lilies are common; rare finds include trilobites and shark teeth.
- **Fort Worth Museum of Science and History** ($).

DENVER

- **Garden of the Gods,** Colorado Springs (free). Get an expansive view of the red sandstone of Pikes Peak from the visitor center's terrace before walking through the iconic Gateway Rocks, one of the most photographed geological formations in America.
- **National Ice Core Laboratory,** Lakewood (free). Schedule a

tour to visit this working lab and storage facility, where scientists study ice from glaciers across the globe, plot Earth's climate history, and predict its future.

- **National Renewable Energy Laboratory,** Golden (free). Consider signing up in advance for a tour that delves into the lab's energy-efficiency research and applications. Education programs teach visitors about reducing their environmental impact.
- **Denver Museum of Nature and Science** ($).

>> **At Garden of the Gods Visitor & Nature Center in Colorado**

DETROIT

- **Ford factory,** Dearborn ($). Get a close-up look at the manufacturing of the F-150 pickup, plus an outline of Ford's history and the steps in building a car.
- **Inland Seas Education Association,** Suttons Bay ($). Activities and classes aboard a tall "Schoolship" sailing on the Great Lakes teach history, geology, ecology and more. Try a night-time

sail with an astronomer, a lesson on building boats, or a guided tour through wetlands.

- **Michigan Science Center** ($).

HOUSTON

- **WaterWorks Education Center,** Humble (free). Schedule an appointment to visit and learn about water science and the city's drinking water supply. Visitors walk through the treatment process from a water drop's perspective.
- **Crocodile Encounter,** Angleton ($). Tours of this zoo allow visitors to observe the big reptiles as they swim, eat and sun themselves.
- **Saint Francis Wolf Sanctuary,** Montgomery (free). Schedule a tour to see rescued residents from afar, and meet a wolfdog, a canine with wolf and domestic dog ancestry.
- **Children's Museum of Houston** ($). Geared toward toddlers to 12-year-olds, the museum makes learning interactive through invention workshops led by profes-

→ The Science Museum of Minnesota

sional "makers," or experienced engineers and designers, and conservation lessons in a native plant garden, for example.

LOS ANGELES

- **Ocean Institute,** Dana Point ($). On weekends, the institute lets the public in to see marine life in its natural habitat and to check out historical maritime artifacts like the 134-foot replica of a 19th-century brig.
- **Griffith Observatory** (free). Look through a high-powered telescope, browse exhibits on space observation, and catch one of the planetarium's shows.
- **Jet Propulsion Laboratory,** Pasadena (free). Explore the NASA and California Institute of Technology lab that constructed the Mars Curiosity rover and America's first satellite.
- **California Science Center** (free).

MIAMI

- **Everglades National Park,** Homestead ($). Rare species such as manatees, American crocodiles and Florida panthers live in the 1.5 million-acre park. Bike, hike, camp, canoe or participate in a ranger-led educational program.
- **Miami International Airport** (free). To learn about aviation, take a tour of the airfield, fire station and other facilities.
- **Patricia and Phillip Frost Museum of Science** ($).

MINNEAPOLIS

- **National Eagle Center,** Wabasha ($). Two floors of exhibits and an observation deck for watching eagles in the wild. Attend a program to learn about the anatomy of eagles and their role in the ecosystem, and see one of the center's five rescued eagles up close.
- **Farmamerica Minnesota Agricultural Interpretive Center,** Waseca ($). June through August, visitors learn about developments in farming and agricultural technology over the past 160 years. The center boasts a country grain elevator and feed mill and a 1930s farmstead.
- **Reptile and Amphibian Discovery Zoo** ($). Specializing in frogs, snakes, lizards, turtles and crocodiles. Attend a feeding on the weekend.
- **Science Museum of Minnesota,** St. Paul ($).

NEW YORK

- **Sony Wonder Technology Lab** (free). Watch email and other Internet traffic zip around a world map. Try programming a robot. You can get a feel for what it's like to perform heart surgery through a haptic controller, which recreates a surgeon's sense of touch.
- **Gateway National Recreation Area** (free). This park spans the coastal areas and waters of three boroughs of New York City and part of New Jersey, offering everything from wildlife sanctuaries and bird-watching to ocean beaches.
- **New York Botanical Garden** ($). Situated on 250 acres in the Bronx, the garden has something in bloom year-round. Find out about plants of the rain forest, desert and more.
- **National Museum of Mathematics** ($). The math problems in this museum look nothing like the ones in textbooks. Go for a ride on a square-wheeled tricycle or tinker with 3-D printers, fractals and electronic paintbrushes.
- **American Museum of Natural History** ($).

PHILADELPHIA

- **Wagner Free Institute of Science** (free). View mounted animals, fossils, shells, bones, an extensive mineral collection and the fossilized skull of the first American saber-toothed tiger, discovered in 1886.
- **Crystal Cave Park,** Kutztown ($). The cave sparkles with calcium crystals and walls of flowstone. Learn about the science behind cave formation.

- **Insectarium** ($). See live and mounted exotic insects and discover insects' influence on human life. Exhibits re-create insect habitats, from trees and tall grasses to kitchen sinks.
- **The Franklin Institute** ($).

PHOENIX

- **Challenger Space Center Arizona,** Peoria ($). Exhibits include interactive displays of the solar

>> Inside the National Air and Space Museum

system, pieces of space shuttles and a planetarium.
- **Desert Botanical Garden** ($). This 140-acre facility offers displays of desert plant life, desert plants' uses to humans and sustainable desert gardening.
- **Arizona Science Center** ($).

PORTLAND, OREGON

- **Zenger Farm** ($). Schedule a visit to observe urban farming and sustainability in action. The farm prides itself on protecting the wetlands nearby by never using chemical fertilizers or pesticides.
- **World Forestry Center** ($). The center's museum has hands-on exhibits featuring tree species from across the globe. Go on a simulated river raft adventure, learn to operate a timberjack harvester, find out how people around the world use forests, and go on a virtual tour of the four types of forests.
- **Oregon Museum of Science and Industry** ($). In the science playground, visitors age 6 and under can use buckets and shovels to experiment with the physics of sand and water. Older visitors can explore the museum's four other halls on life science, technology and more.

SAN FRANCISCO

- **San Francisco Maritime National Historical Park** ($). This park serves up history, science and engineering through a re-creation of the city waterfront over the decades, early-morning birdwatching led by rangers, and demonstrations of 19th-century technology.
- **Cable Car Museum** (free). Learn about the engineering of centuries past by viewing old cable cars, detailed models and such mechanical devices as grips, tracks, cables and brakes.
- **California Academy of Sciences** ($).

SEATTLE

- **Future of Flight Aviation Center and Boeing Tour,** Mukilteo ($). Visit a commercial jet assembly plant and take a factory tour to see how aircraft are constructed. In the aviation center, learn about airplane design, flight systems, propulsion and the future of aviation.
- **Stonerose Interpretive Center & Eocene Fossil Site,** Republic ($). The fossil site is a dried lake whose fish, plants and insects have been preserved for 50 million years by volcanic ash. Use hammers and chisels to dig up fossils that experts will help you identify.
- **Pacific Science Center** ($).

WASHINGTON, D.C.

- **National Arboretum** (free). The 15 gardens and collections include extensive displays of azaleas, herbs and dogwoods and a bonsai museum.
- **The National Zoo** (free). Known for its giant pandas, the National Zoo houses some 1,800 animals representing 300 species.
- **National Inventors Hall of Fame,** Alexandria, Virginia (free). On the same campus as the U.S. Patent and Trademark Office; displays the technological accomplishments of over 500 inductees.
- **National Air and Space Museum** (free). In two locations in Washington and suburban Virginia, it contains the country's largest collection of aviation and space artifacts, including the Wright brothers' plane and the Apollo 11 command module. See rocks from the moon here, too.
- **National Museum of Natural History** (free).
- **Maryland Science Center,** Baltimore ($). ●

Curl Up With a Great Read

Whether your child wants to learn more about animals or explore the life of a famous female scientist, your local library or bookstore can help. Here's just a sampling, for readers of all ages, of the many great books and series that can ignite your son's or daughter's inner scientist:

Preschool

- **Bedtime Math** (series) by Laura Overdeck (Feiwel & Friends)
- **Big Blue Whale** by Nicola Davies (Candlewick)
- **The Cat in the Hat Knows a Lot About That!** (series; Random House)
- **Let's-Read-and-Find-Out Science** (series; HarperCollins)

- **The Most Magnificent Thing** by Ashley Spires (Kids Can Press)
- **The Watcher: Jane Goodall's Life with the Chimps** by Jeanette Winter (Schwartz & Wade)
- **Young Frank, Architect** by Frank Viva (The Museum of Modern Art)

Elementary Grades

- **Alien in My Pocket** (series) by Nate Ball (HarperCollins)
- **Batman Science** (series; Capstone)
- **The Boy Who Loved Math: The Improbable Life of Paul Erdos** by Deborah Heiligman (Roaring Brook Press)
- **Citizen Scientists: Be a Part of Scientific Discovery from Your Own Backyard** by Loree Griffin Burns (Square Fish)
- **Engineer Through the Year: 20 Turnkey STEM Projects to Intrigue, Inspire & Challenge** by Sandi Reyes (Crystal Springs Books)

- **Eye to Eye: How Animals See the World** by Steve Jenkins (Houghton Mifflin Harcourt)
- **Fun with Nature: Take-Along Guide** by Mel Boring, Diane Burns and Leslie Dendy (Cooper Square Publishing)

- **Girls Think of Everything: Stories of Ingenious Inventions by Women** by Catherine Thimmesh (HMH Books for Young Readers)
- **The Grapes of Math** by Greg Tang (Scholastic)
- **Growing Patterns: Fibonacci Numbers in Nature** by Sarah C. Campbell (Boyds Mills Press)
- **The Kid's Book of Simple Everyday Science** by Kelly Doudna (Mighty Media)
- **Kitchen Science Lab for Kids: 52 Family-Friendly Experiments from Around the House** by Liz Lee Heinecke (Quarry Books)
- **Life in the Ocean: The Story of Oceanographer**

Sylvia Earle by Claire A. Nivola (Farrar, Straus and Giroux)
- **The Magic School Bus** (series) by Joanna Cole (Scholastic)
- **Math Curse** by Jon Scieszka (Viking)
- **Nat Geo Wild Animal Atlas: Earth's Astonishing Animals and Where They Live** (National Geographic Kids)
- **Newton and Me** by Lynne Mayer (Sylvan Dell Publishing)
- **Numbed!** by David Lubar (Millbrook Press)
- **Rosie Revere, Engineer** by Andrea Beaty (Abrams Books for Young Readers)
- **Sally Ride: Life on a Mission** by Sue Macy (Aladdin)
- **Seymour Simon's Extreme Earth Records** by Seymour Simon (Chronicle Books)
- **Star Stuff: Carl Sagan and the Mysteries of the Cosmos** by Stephanie Roth Sisson (Roaring Brook Press)
- **Super Nature Encyclopedia** (DK)
- **The Ultimate Book of Science** (Oxford University Press)
- **What Color Is My World? The Lost History of African-American Inventors** by Kareem Abdul-Jabbar and Raymond Obstfeld (Candlewick)
- **Who Says Women Can't Be Doctors? The Story of Elizabeth Blackwell** by Tanya Lee Stone (Henry Holt and Co.)
- **Your Backyard Is Wild!** by Jeff Corwin (Puffin Books)

Grades 6-8

- **Are We Alone? Scientists Search for Life in Space** by Gloria Skurzynski (National Geographic Children's Books)

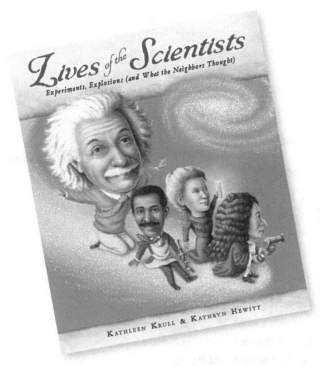

- **Case Closed? Nine Mysteries Unlocked by Modern Science** by Susan Hughes (Kids Can Press)
- **Headstrong: 52 Women Who Changed Science – and the World** by Rachel Swaby (Broadway Books)
- **How Come? Every Kid's Science Questions Explained** by Kathy Wollard (Workman Publishing Company)

- **Lives of the Scientists: Experiments, Explosions (and What the Neighbors Thought)** by Kathleen Krull (Harcourt)
- **Math Doesn't Suck: How to Survive Middle School Math Without Losing Your Mind or Breaking a Nail** by Danica McKellar (Plume)
- **The Mighty Mars Rovers: The Incredible Adventures of Spirit and Opportunity** by Elizabeth Rusch (Houghton Mifflin Harcourt)

- **Nick and Tesla** (series) by "Science Bob" Pflugfelder and Steve Hockensmith (Quirk Books)
- **One Minute Mysteries** (series) by Eric Yoder and Natalie Yoder (Science, Naturally!)
- **Scientists in the Field** (series; Houghton Mifflin Harcourt)
- **Secrets of Mental Math: The Mathemagician's Guide to Lightning Calculation and Amazing Math Tricks** by Arthur Benjamin and Michael Shermer (Three Rivers Press)
- **Temple Grandin: How the Girl Who Loved Cows Embraced Autism and Changed the World** by Sy Montgomery (Houghton Mifflin Harcourt)

Grades 9-12

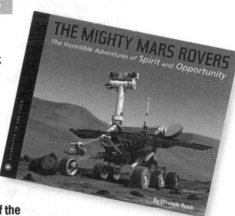

- **The Cartoon Guide to Physics** by Larry Gonick (Harper Perennial)
- **Death by Black Hole: And Other Cosmic Quandaries** by Neil deGrasse Tyson (W. W. Norton & Company)
- **E=mc2: A Biography of the World's Most Famous Equation** by David Bodanis (Berkley Publishing Group)
- **The Ghost Map: The Story of London's Most Terrifying Epidemic – and How It Changed Science, Cities, and the Modern World** by Steven Johnson (Riverhead Books)
- **Rocket Boys: A Memoir** by Homer H. Hickam Jr. (Delacorte Press)
- **A Short History of Nearly Everything** by Bill Bryson (Broadway Books)
- **What If? Serious Scientific Answers to Absurd Hypothetical Questions** by Randall Munroe (Houghton Mifflin Harcourt) ●

Standout Science Shows

From animated series to YouTube videos, families can find a wide range of engaging TV and video content to excite kids about science, technology, engineering and math. The best of these shows not only teach children basic concepts that should help them excel in school, but also guide them toward becoming explorers and experiencing the thrill of discovery. Here's a taste of what's available:

For the younger set

Anyone with a toddler or preschooler is likely familiar with PBS Kids, whose packed menu of science- and math-related television shows is led by **Sesame Street,** the industry standard-bearer when it comes to preschool educational programming. The franchise known for using Big Bird, Elmo and the rest of the gang to help foster early literacy and understanding of numbers is branching out. Its parent organization, Sesame Workshop, now hosts **Little Discoverers,** a destination site featuring videos on such topics as why objects float and how to measure, starring Sesame characters (search "Little Discoverers" or visit sesame.org).

Other PBS Kids offerings include **Sid the Science Kid**, whose hero, a preschool-age aspiring scientist, investigates important questions like "How do rainbows form?" and "How do computers work?" Young dinosaur lovers can hop aboard **Dinosaur Train** to learn about natural history and paleontology. The network's **Wild Kratts** is hosted by real-life brothers Chris and Martin Kratt, animated versions of whom make fantastical visits to animals in their habitats – donning special suits to fly with peregrine falcons, for instance, or having a contest to see if elephants are stronger than rhinos.

Want your preschooler to develop math skills early? Try **Peg + Cat** (also from PBS Kids), which follows a young girl and her feline friend as they use math to tackle all sorts of problems, from helping pirates divide up their fruit fairly to using a diagram to escape from a baby T. rex in a prehistoric

>> "SciGirls" takes a hands-on approach to science.

forest. Similarly, Nick Jr.'s **Blaze and the Monster Machines**, an animated series about racing trucks, features characters who use math concepts to calculate speed, say, or turn a truck into a makeshift wrecking ball to clear a tunnel obstruction.

For K-8 students

If your child is fascinated by tinkering and building, drop in on the Emmy award-winning web-based **Design Squad Nation** (pbskids.org/design-squad). The site presents engineering puzzles and details step-by-step solutions that viewers can try repeating at home. One example: using air from a deflating balloon, transmitted through a straw, to power forward a toy car the assemblage is attached to. The site also houses videos explaining mysteries like the engineering behind Slinky toys and highlighting the stories of role models like volunteers from Engineers Without Borders helping to build a dam for an impoverished community in Cambodia.

SciGirls, now on PBS Kids, also emphasizes hands-on learning. The show follows real middle-school-age girls as they learn about how ornithologists track birds, say, or examine an artificial oyster reef in the Chesapeake Bay with the help of a robotic underwater vehicle.

Amateur sleuths who want to polish their math and detective skills can try out the quirky **Odd Squad,** another

» **Agents Olive (left) and Otto use math to solve offbeat cases on "Odd Squad."**

PBS Kids program. The show follows "government agents" Olive and Otto as they take on such unusual cases as stopping missing zeroes from destroying their town and finding the thief who's vandalizing symmetrical shapes. And don't forget **Bill Nye the Science Guy.** Originally airing on PBS, the popular series for years put forth an engaging mix of science, music and fun while tackling everything from the ecosystems of wetlands to the body's biggest organ: the skin. Nye's old shows and new content may be found on YouTube and his own website (billnye.com).

For a more sophisticated take

For older kids interested in a deeper dive into topics, **Modern Marvels**

(History Channel) looks at engineering, archaeological and scientific feats around the world, from the evolution of waterproofing technology to the massive engineering effort underway to expand the Panama Canal.

How It's Made (Science Channel) takes a soup-to-nuts look at the manufacturing process for every imaginable product, from handcrafted skis to wind turbines and industrial mixers. Science Channel's **Build It Bigger** spotlights engineers and architects taking on complex tasks like building one of the world's tallest skyscrapers or designing hurricane-proof homes. And the long-running science series **NOVA** (PBS) offers topics for virtually every taste, ranging from the ancient secrets of Stonehenge to the engineering challenges behind the Hubble Space Telescope. ●

Your Daughter, the Scientist

When it comes to exciting kids about STEM, girls sometimes require special attention. That's because they learn differently than boys do, and they can be far more discouraged by failure. A 2012 report by the Girl Scout Research Institute found that 74 percent of girls start off interested in STEM subjects; still, somewhere in middle school, interest fades. Today, only 12 percent of engineers in the U.S. are women.

Your encouragement is crucial to improving these numbers. "Parents don't have to be the experts," says Linda Kekelis, CEO of Techbridge, an Oakland, California, organization that offers programs and activities to inspire girls in STEM. "They just have to be supportive." Sometimes it's as simple as giving your daughter time away from chores and schoolwork and providing transportation to an after-school program, watching a TV science program like PBS's "SciGirls" or "Design Squad Nation" with her, or setting aside a special area where she can experiment.

Today, we face complex challenges in medicine, energy, the environment, transportation and agriculture. Your daughter can provide the solutions – with a helping hand from you:

Start early

It's never too early to fight STEM stereotypes, so seek out games, toys and books that show that girls can be environmental scientists, biomedical engineers and mobile app developers too, says

Dara Richardson-Heron, CEO of YWCA USA. For example, young readers can be inspired by books such as "Girls Think of Everything: Stories of Ingenious Inventions by Women."

Learn after school

Not all education takes place in the classroom. Check out your local library, science or natural history museum or nearby college or university for after-school enrichment programs as well as community organizations and clubs. The Girl Scouts, for example, have proficiency badges in innovation, science and technology and digital arts; and local YWCAs provide activities focused on engineering, robotics, computer science and biotechnology.

Find role models

Many girls want to make the world a better place. Show them that women make a difference. They solve problems; they make scientific discoveries; they protect the environment; they explore space. Some of the famous women in STEM include Marie Curie, who discovered radium and helped create X-rays; Shirley Ann Jackson, president of Rensselaer Polytechnic Institute and the first African-American woman to earn a doctorate from the Massachusetts Institute of Technology; and Ellen Ochoa, an electrical engineer and the first Hispanic female astronaut.

Closer to home, ask your child's teacher or guidance counselor, church members, librarians or museum curators for names of local role models. If there is a college near you, students themselves can become mentors. "It's sometimes easier for young girls to relate to someone who is just a step beyond them," says Kekelis.

Bring it home

How does a grocery scanner work? Why doesn't a bridge fall down? Where do bees go in the winter? Point out how science, math and engineering are part of daily life. Talk about science and technology at the kitchen table. Do projects with your daughter. Your local librarian can recommend books such as "Citizen Scientists: Be a Part of Scientific Discovery from Your Own Backyard" and "Kitchen Science Lab for Kids: 52 Family Friendly Experiments from Around the House" to give you some ideas.

Build confidence

Girls do just as well in math and science as boys, but their confidence is lower. Help your daughter get beyond the fear of failure by telling her that mistakes and risk are all about learning. "Praise the effort, the process," suggests Kekelis. Let her know she does not have to have all the answers. Encourage your daughter to do hands-on projects; get messy; work on something, test the design and then rework it.

Try the team approach

Working in teams can be particularly valuable, says David Etzwiler, CEO of the Siemens Foundation, which sponsors an annual science competition that high school students can enter either as individuals or as groups. "We see a lot of girls coming forward in teams," Etzwiler says. "It's important to encourage girls to be risk-takers and to be willing to be wrong. They really like to team up to take those risks, to exchange ideas and work together."

Embolden and encourage

Finally, be mindful of the messages you're passing along to your daughter. If your child is floundering in one STEM topic, try gently pointing out alternatives where she might excel. When children struggle "with one type of math or one science class, they decide they're unable to do any of it," says Erika Ebbel Angle, CEO of Science from Scientists, a Boston group that brings scientists into schools. "Tell them that just because they don't understand one subject they shouldn't give up on it." An open mind is key. "One of the biggest things is just to make sure that the girls try everything: robotics, electronics, 3D imagery, game design, mobile apps," says Kimberly Bryant, founder and executive director of San Francisco-based Black Girls Code. ●

will.i.am
Musician; STEM advocate

Will.i.am, the Grammy Award-winning musician and Black Eyed Peas frontman, is an enthusiastic STEM advocate, inspiring students through his i.am.angel Foundation and other initiatives.

What sparked your interest in STEM?

One [thing] was a [2010] movie by the name of "Waiting for 'Superman'" that talks about the education system in America and how poorly it performs. In particular, my neighborhood [in Los Angeles] that I come from was featured in that movie. Superman, a fictitious character, is supposed to solve real problems. STEM, to me, is the solution for schools and neighborhoods like mine.

How have you worked to improve STEM education?

To help solve the problems and the riddles that plague my community and the communities like it … we created this cross-disciplinary, transformative, project-based-learning curriculum that kids do after school. Our kids had a 0.74 GPA – just failing beyond failing – and now they have 3.4s, 4.0s. If you're living in the hood and you're surviving, what incentive do you give kids? We say, "Let's get on track to go to college, learn the skill set, so not only are you looking for a job when you get out of college, you can create jobs."

Do you have any other thoughts about changing the culture around STEM?

When I was going to elementary school, we had science in our school. I went to Brentwood Science Magnet. We had science class, oceanography lab, physics and computer labs. And then, somewhere in the '90s, they started cutting budgets. They took music out of schools; they took science out of schools. What built America was STEM. It was companies like Ford – that's engineering. It was H-1B visas – we were able to bring people from other countries. Thank God that it's a subject making its way to popular culture because for some reason popular culture forgot the importance of science, technology, engineering and mathematics.

How can parents help?

At the [2014 White House] Maker Faire, there was this beautiful robot. Who built this? Two little girls. [I asked,] "How old are you?" "I'm 14." "I'm 12. I'm her little sister." I'm like, "So who designed it?" Here walks the dad: "I helped them design it. … Every weekend, me and my girls go in the garage, and we start building robots." Wow. If it was a dad and a son, that sounds pretty obvious. But two girls and a pop? So that is an amazing story to see parents and kids – especially girls – building robots. They're going to take that skill set with them to high school, and then, when they graduate high school, they're probably going to go to MIT or Stanford. Then they're going to get a job at Lockheed Martin or Boeing or the Department of Defense. Amazing things. ●

Smart Activities
for Your Kid's Spare Time

>> **Virtually every community offers children opportunities to dive into STEM – many free of charge. These after-school, weekend or summer programs give kids the chance to roll up their sleeves and do all sorts of hands-on experimentation, from developing computer and coding skills and building robots to exploring local rivers and forests and wildlife. Here are just a few of the resources you can tap into:**

- **After-School All-Stars** (afterschoolallstars.org)

- **Boy Scouts of America** (scouting.org)

- **Boys & Girls Clubs of America** (bgca.org)

- **Camp Fire** (campfire.org)

- **4-H** (4-h.org)

- **Girl Scouts of the USA** (girlscouts.org)

- **Girls Inc.** (girlsinc.org)

- **National Association of Police Athletic/Activities League Inc.** (nationalpal.org)

- **YMCA** (ymca.net)

- **YWCA** (ywca.org)

OTHER LOCAL RESOURCES

- **Local libraries.** Many have ramped up after-school STEM programs.
- **Museums.** Various science and natural history museums, such as the American Museum of Natural History in New York and the Science Museum of Minnesota in St. Paul, offer after-school programs at no cost.

- **Parks and Recreation Departments.** Look here for nature hikes, stargazing, animal lectures and more.
- **21st Century Community Learning Centers.** Schools nationwide receive government funds to provide after-school programs to help students reach state and local achievement standards for core subjects like reading and math.

MyStory →

Durrell Hightower
Technician, LA Freightliner

While still in high school, I found myself in charge of my younger sister and brother because of family problems. To support them, I finished high school, got a job at McDonald's and became an assistant manager at 19. Eventually, I realized I wanted something that paid better, was more fun and that I enjoyed doing. In school, I'd always loved math and science. I also loved to fix things, to tinker and figure out how things work, especially cars. But the vehicles now all have computer systems and electronic control modules, and you need to be trained on them. I did my research and found a school, Universal Technical Institute, that had good relationships with major auto companies and a good job placement record. My UTI program included studies of electrical components, circuit routing, data link systems, fault code diagnosis, and computer-based software for system troubleshooting.

I eventually earned two certificates: one for diesel and industrial technologies and another specifically for Daimler trucks. The certificates made it possible for me to get a job at a dealership that sells and services Daimler trucks. After just two years, I've become a supervisor. Eventually, I hope to manage my own operation. But it all starts with making sure you have the right technical skills to prepare you for the job you want and that you're with a school that will actually help you get that job. I love what I do and will continue to grow. As technologies improve, so will I. **"**

Patty Legaspi
Senior Test Engineering Manager, Google

I was the kind of student who always hung around teachers. My parents didn't go to college, but they encouraged me to reach farther than they had. Going to school in Oakland, California, I always found people to connect with who could offer me guidance.

I was fascinated by computers. In school, I took an animation class and learned some basic programs. I knew then I would love a career in computers but didn't know how to get there at first. Fortunately, I had a high school French teacher who really encouraged me. Whenever I came to her with a question, she asked me questions in return and made me think. She talked me through my fears about paying for college. She told me to look for scholarships. I would find the money. She also suggested I visit some campuses, which is how I realized that a small college where I could get more personalized attention was a better fit. I ended up at Mills College in Oakland, where I got a computer science degree.

I had another great adviser at Mills who urged me to accept a summer research project in the Bay Area. She knew I wanted to stay near my family, so this is where, she said, I should build my professional network. This same professor later shared my résumé with a recruiter from a young company called Google. Today, I head a team that manages how Google's Chrome OS interfaces with different devices. I remember as a child my parents urged me to think big and as Google has grown, I have too. **"**

Suc

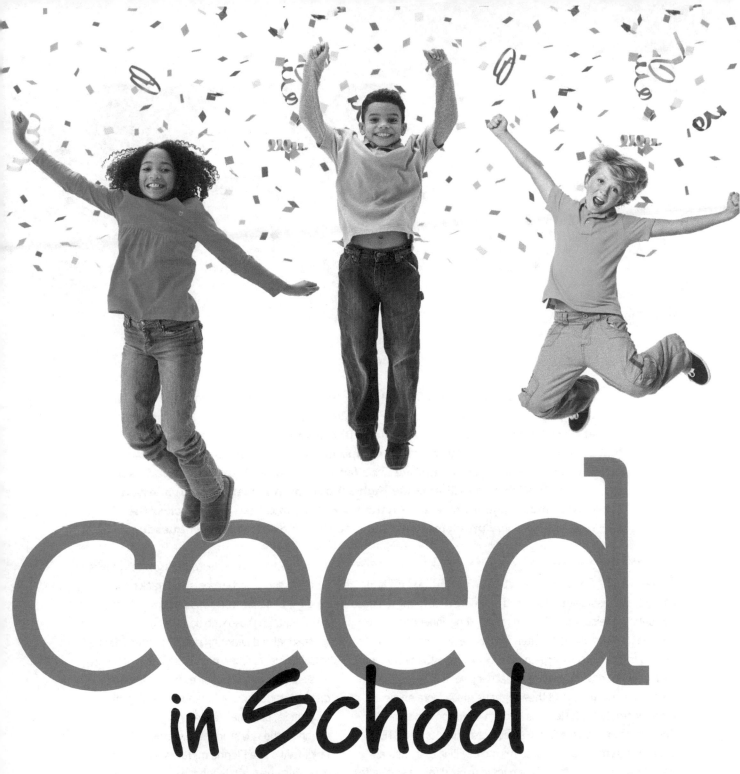

ceed in School

GETTY IMAGES (5)

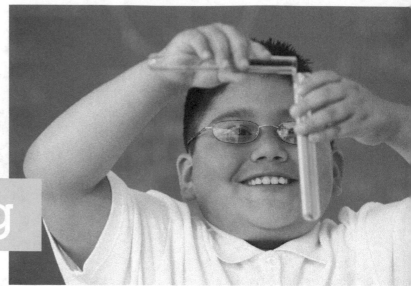

Learning by Doing

At Jennie Dean Elementary School in Virginia, first-graders learn about design (and physics!) by figuring out how to build a boat out of tinfoil that can hold at least 10 pennies for three minutes without sinking. At Plaquemine High School in Louisiana, students in a computer class apply programming lessons they've learned to create mobile apps (and earn money if they manage to sell them). At Michigan's Clintondale High School, math students watch a lesson on advanced algebra by video on their own time and then go into class the next day to solve problems while the teacher stands by to answer questions.

Practical and real-world learning is in,

as rigid, old-fashioned methods of instruction are being replaced throughout the country. Educators have long understood that the one-size-fits-all model often fails to reach students who learn best through other methods. For example, many students might start to daydream if their teacher gives a dry presentation on how gravity works, but they might be much more engaged if they learn the same lessons through hands-on experiments. The growing national effort to steer more students into STEM subjects has made school systems more willing to try new approaches. Here are some of the ones your child might encounter:

In personalized learning, sometimes called

student-centered learning, teachers try to figure out how individual students learn best and then give them more customized lessons that play to their strengths. At Fulton High School in Tennessee, students can choose whether to read a book or watch videos while learning chemistry, biology and environmental science. Personalized learning also often means students set their own goals and track their progress because research has shown that they do better than those who don't. At the Academy of Personalized Learning public charter school in California, teacher, parent and student meetings are held at least once every 20 days to discuss student progress and to revise goals. Elsewhere, personalized learning may involve changing the way students are grouped and taught. At some Florida middle schools, instead of pairing one math teacher with 20 students, there are three math teachers for 60 students. Why? More teachers equals more chances for students to find a particular instructor they click with, and vice versa. Students also often work independently on computers and form groups of their own to discuss topics they didn't understand. New York City's School of One uses a computer algorithm to create each kid's lesson plan "playlist" for the next day based on how well he or she did on tests the day before.

In flipped classrooms, a lot of
the "teaching" happens at home in place of traditional homework assignments. Students learn material on their own, usually with the help of teacher-created videos or other media, and participate in online discussions, freeing up class time to work on problems, activities and projects. At Byron High School in Minnesota, students complete math lessons online, at home, and use class time to solve problems, ask questions and debate their answers with other classmates, while the teacher is on the spot to help as needed.

Online discussions, held out-
side of class, are also popular because they avoid some problems of face-to-face ones. In class, students may be nervous about raising their hands or need more time to formulate an answer. But online, they can post an answer when they're ready, and the teacher and other students can respond. Plus, teachers can read along and see what misunderstandings are occurring, and give those extra attention in class.

Blended learning focuses on
cutting down on the amount of teacher lectures in favor of more interactivity. Students do some learning online, typically using apps (like a graphing calculator or iPocket Draw) and videos or audio recordings to absorb various ideas and concepts or to carry out certain tasks. Digital content has an added advantage: It tends to be more up-to-date than textbooks and gives kids a more dynamic, interactive experience.

Project-based learning
"Problem-solving" assignments in school traditionally have involved students being taught how to get to a single right answer – not the greatest preparation for the real world, where problems are rarely well-defined and there can be many potential solutions. Project-based learning tries to bridge this gap by giving students actual problems to solve, often teamed up with their classmates, with the teacher serving as a guide. For example, last year, fifth-graders at Northwood Elementary in Georgia studied a local zoo and came up with ways to redesign it with different types of animals and plants. They presented their proposals in class. In another school, students might be asked to serve as geologists, studying the raw materials available in their hometown to determine which ones could be used to build or strengthen local roads and buildings, and then present their findings to the town council.

As teachers increasingly adopt more flexible ways to teach math and science, more kids are learning not just to enjoy these subjects, but to excel at them. ●

THE NEW YARDSTICKS

Many states have adopted the Common Core State Standards with the goal of better preparing young people for college and careers. The academic standards establish consistent benchmarks for what skills K-12 students should have mastered in math and reading by the end of each grade. To learn more about Common Core, visit corestandards.org.

About a dozen states have similarly approved the Next Generation Science Standards, which aim to update and standardize science curricula. For more information, visit nextgenscience.org.

Turn School Setbacks Into Success

In middle school, my science-loving daughter failed her first big math test and announced she "hated" math. School failures can be overwhelming for kids. Students who fail over and over may start to believe they don't have the skills to succeed. Or, they may be so afraid to fail, they don't even try. But making mistakes, messing up, being wrong – that's all just a part of life.

Engineers routinely "test to failure," meaning they keep testing materials or designs until they figure out what will work best in new technologies. The Museum of Science in Boston works with thousands of educators to teach this engineering mindset to students, who can then apply it to many fields of study. Here are three effective approaches that engineers use that parents can adopt, too.

Everyone fails. One way to help kids manage their frustration is by reminding them that even successful people make mistakes or hit a dead end sometimes. You might share a time when you failed at something on your first try. Describe your feelings – how you felt embarrassed or upset – but also talk about what you learned and what you did to be successful eventually. You might also share examples of famous people who failed big, like Theodor Geisel, aka Dr. Seuss, whose first book was rejected by 27 publishers; Abraham Lincoln, who lost numerous bids for public office before winning the presidency; or Oprah Winfrey, who was fired from her first job as a news anchor by a boss who told her she was "unfit for TV."

Try again. Your child didn't fail. His plan, his work, his performance failed. When the bridge collapses in a design project or your child's presentation doesn't go well, for example, try framing these setbacks as design or planning failures. Resist the temptation to criticize. Be disappointed along with your child, not at her. You might say: "This assignment (or test) didn't turn out like you expected. I know you're frustrated. But you can try again. I know you're capable of succeeding."

Let's troubleshoot it. When your child has a setback, diagnose the problem as engineers do. What went wrong and why? Encourage your child to brainstorm ways he might approach things again differently. You'll send the implicit message that he can solve problems on his own. With freedom to fail, kids engage more deeply in learning. We constantly hear from teachers that their students beg for more time to take their projects home so they can keep improving them. When you help kids hang in there when things get tough, apply what they've learned from failure and try again – you're preparing them for school and for life.

By the way, my daughter's teacher allowed her to take a makeup test. After studying and taking practice exams for two weeks, her hard work paid off – she earned a 96. ●

By Christine M. Cunningham, vice president at the Museum of Science, Boston, and director of Engineering is Elementary, a project of the National Center for Technological Literacy.

Psst! The Secret to Sticking With Math

Middle school can be a scary time for kids. Hormones are raging and math, which once seemed easy and related to the real world (like dividing a pie into equal parts or counting by tens), has suddenly become abstract. Students find themselves wrestling with dividing fractions and solving equations and can begin to lose confidence. Here are some ways you can help your child work through these all-too-common slumps:

Engage your child

Often, just reading a book beside your child during homework time can have a calming effect. If your son can't figure out an assignment, don't do the work but ask questions to zero in on problem areas: "What part do you understand?" "What don't you understand?" "Can you make a drawing to explain your thinking?" Sometimes rewording an assignment question can help, too. If he is getting frustrated, then suggest he go shoot baskets for 20 minutes, perhaps, and then come back. If something is still unclear, then he can write down a question for the teacher (most will be very pleased to help), who then can more easily provide the needed support. You can also investigate websites like Khan Academy, which has terrific video tutorials (that parents can take too!), practice problems and articles covering many math concepts.

Be positive

If your child is struggling in class, keep encouraging her that she can do the work. Praise her effort, not her ability. Contrary to popular belief, there is no math gene. Learning is a matter of time, effort and practice. When kids believe they can succeed, their minds are more open to learning. Nurture their curiosity while providing support and encouragement.

Learn from mistakes

Part of learning math also comes by working through failure. The better you get, the more your confidence builds. Don't panic over a poor test result. Most teachers are willing to raise a grade if a child does extra work and later demonstrates mastery of the material that tripped him up on the test.

Go beyond the classroom

The Internet has so many learning opportunities. If your child wants to find out more about imaginary numbers, say, suggest she Google the phrase and then tell you what she discovered. Many extracurricular activities are also available to apply math concepts, like robotics clubs or summer STEM camps.

And you can always talk about math in everyday life, whether it's comparing cellphone plans or looking at how architects use geometric shapes in building structures, like skyscrapers or a neighbor's home. Construction toys like Legos and games like checkers or Minecraft are also great vehicles for conveying math concepts and the problem-solving process. ●

By Jane Porath, a board member of the National Council of Teachers of Mathematics who teaches math at East Traverse Middle School in Michigan.

Gear Up for the Science Fair

» For these competitions, students design and conduct their own science experiments or solve engineering problems. They then prepare an oral and visual presentation on their findings, generally with the aid of a poster board. Winners may go on to participate in larger state, regional or national competitions. Washington's 2015 D.C. STEM Fair (below and right) drew local middle and high school students, who studied topics ranging from how music affects blood pressure to how electromagnetic radiation from cellphone calls can be reduced.

» Students discuss the results of their projects.

» A 12th grader pitches her study of seed germination to a judge.

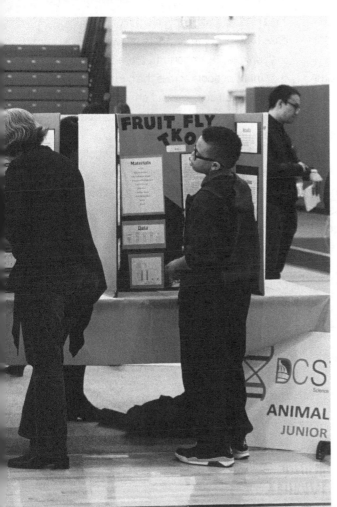

Checklist for a Winning Entry

The head of the California State Science Fair shares his tips for students on designing a smart presentation:

✔ **Find a mentor.** Try a teacher first, but otherwise it can be another student with fair experience or an adult with a science or engineering background.

✔ **Google "winning science fair projects"** online for inspiration, then come up with your own idea.

✔ **Develop a project timeline** that includes tasks like: conducting background research, creating a testable question for a scientific experiment or a solution for an engineering problem, writing a report, and designing the presentation.

✔ **Keep a journal** to organize, date, label and record your observations, procedures, drawings and data you collect.

✔ **Design a mock presentation board** with text, pictures, labels and graphs before doing the real one.

✔ **Lay out the project content** in a clear, logical manner, using subheadings, bullet points and short, descriptive sentences. Proofread everything before attaching to your fair board.

✔ **Choose easy-to-read lettering or type** that can be read from four feet away.

✔ **Make sure your oral presentation** has a clear and concise introduction, explanatory section and conclusion.

✔ **Practice a lot** with whomever will listen, including family and friends.

✔ **Bring a "first aid" kit** with supplies such as tape, glue, a stapler and scissors to the fair to make any last-minute fixes to your display.

✔ **Look and be professional.** Act confident, enthusiastic and, most of all, have fun!

Ron Rohovit is Deputy Director of Education at the California Science Center, which hosts the annual California State Science Fair.

01010010 01100101 01100001 01100100 01111001 00100000 01010011 01100101 01110100 0
01110100 00100000 01000011 01101111 01100100 01100101 01010010 01
01100001 01100100 01111001 00100000 01010011 01100101 01110100 00
100100 01100101 01010010 01100101 01100001 01
01111001 00100000 01010011 01100101 01110100 00100000 01000011 0
01100100 01100101 01010010 01100101 01100001 01100100 01111001 00
01010011 01100101 01110100 00100000 01000011 01101111 01100100 01
01010010 01100101 01100001 01100100 01111001 00100000 01010011 0
01110100 00100000 01000011 01101111 01100100 01100101 01010010 01

Ready, Set, Code!

From paying bills online to monitoring your health on a mobile device, computers are a part of everyday life. In the future, understanding the complex details of how they work will be an essential skill for your child, no matter his or her career path. Many of the hottest jobs in the coming years require a deep knowledge of computer science, and careers like cybersecurity analyst and software developer offer fun, hands-on opportunities to design video games, for instance, or to build programs that help protect precious information for consumers.

Of course, not everyone is a born computer programmer, but educators, government leaders and many others say that anyone can benefit from trying a little computer science. Even President Barack Obama has tried his hand at coding, noting how important it is that "young people are familiar not just with how to play a video game, but how to create a video game." Many other famous figures, from NBA all-star Chris Bosh to musician will.i.am to actor Ashton Kutcher, have championed the cause of computing. As technology advances, just about any job, from health and education to art and manufacturing, will require at least some work with computers. Someone who boasts basic coding skills might be looked at more favorably by a college admissions officer or hiring manager.

Plus, computer science isn't simply about staring at numbers and letters on a screen all day. It requires building valuable skills in problem-solving, critical and analytical thinking, design, collaboration and other abilities

that have a wide-reaching impact. "Learning to write programs stretches your mind, and helps you think better, creates a way of thinking about things that I think is helpful in all domains," Microsoft founder Bill Gates has said. (His first coding project was building a computer program to play tic-tac-toe.)

While a growing number of people are making the case that teaching computer science to youngsters is important, there aren't that many schools across the country offering such courses. But children can practice coding and other skills at home through free websites like Code. org and Codecademy.com. Afterschool or summer programs from groups like Girls Who Code, Black Girls Code and CoderDojo can also give students some hands-on experience with computing, not to mention the chance to meet and work with mentors in the field. Many colleges and companies are making a special effort to recruit women, African-Americans and others who are underrepresented in computer science fields, so scholarships and job opportunities are worth exploring for those individuals. ●

Handy Sources of Homework Help

Giving your kids a hand with their homework might be as easy as pointing them to online assistance. Some of these resources present great opportunities for you to sit and learn together. Others are large databases that you can search for information on whatever concept is challenging your child.

Math & Science

- **Ask Dr. Math** (mathforum.org/dr.math). Managed by the Drexel University School of Education, this site features questions from student visitors arranged by topic and grade level. You can also ask a new question (as long as it's not a request to solve a specific problem).
- **The Calculus Lifesaver** (press.princeton. edu/video/banner). Watch videos of former Princeton prof Adrian Banner walk students through his guide: "The Calculus Lifesaver: All the Tools You Need to Excel at Calculus."

- **Discovery Education** (discoveryeducation.com). Videos and graphics break down aspects of earth science, physics, life science and more. A math section includes explanatory material as well as a service, WebMath, that lets you search for guidance.
- **Harcourt Animated Math Glossary** (harcourtschool.com/glossary/math2). Find illustrated definitions of terms and concepts kindergarteners through sixth graders should know.
- **Interactivate** (shodor.org/interactivate). For science and math students in middle and high

school. The dictionary tool is especially useful for homework challenges.

- **MathWorld Classroom** (mathworld.wolfram.com/classroom). Defines terms and concepts for amateur mathematicians in middle and high school and their parents.
- **NSF Classroom Resources** (nsf.gov/news/classroom). The National Science Foundation has compiled a wealth of resources from across the web. Topics range from the Antarctic to chemistry and engineering.
- **Physics To Go** (compadre.org/informal). This site calls itself "an online monthly mini-magazine and a collection of more than 1,000 websites" where student physicists can dig for information in webcasts, online articles and exhibits. Browse through subjects such as electricity, general physics and light; or search to find a site that can help you.

Computing & Engineering

- **Code.org.** Tutorials suitable for elementary to high-school kids (and nontechie parents).
- **Codeacademy** (codeacademy.com). Teaches skills from building a website to programming languages. Students can use the site to supplement their computer science courses.
- **How Stuff Works** (howstuffworks.com). The animals, health, science and technology sections on this Webby Award-winning site offer articles for students in middle school and up.
- **MIT's Science Out Loud** (k12videos.mit.edu). The online video series is created by MIT students and covers topics from the physics of skydiving and invisibility cloaks to how computers compute.
- **NASA Education** (nasa.gov). Features articles,

videos and podcasts on everything from the Hubble telescope to the latest missions.

All Subjects

- **CK-12** (ck12.org). Explanations and interactive lessons on English, history and all STEM subjects from elementary math and statistics to astronomy and engineering.
- **Cliffs Notes** (cliffsnotes.com). Get online help with middle- and high-school math and science as well as history and English. Study aids are arranged by topic and level and contain definitions of terms and examples.
- **Infoplease.com.** Features a homework center of explanatory material by subject, plus an encyclopedia, an atlas and biographies.
- **Khan Academy** (khanacademy.org). The wildly popular Khan Academy offers interactive lessons on arts and humanities topics as well as math and science.
- **Kids.gov.** The government portal offers a wealth of resources for students in kindergarten through eighth grade from the EPA, NOAA, Smithsonian and other agencies and organizations.
- **Shmoop** (shmoop.com). This Webby Award-winning site provides guidance in pre-algebra through calculus, biology and chemistry for middle- to high-school students and parents. ●

YOUR LIBRARY

The Web is a great place to get free homework help, but it's not your only option. Public libraries house shelves of information on the sciences, of course, and they provide easy access to computers. Many also host STEM programs for kids. To find a library near you, search usa.gov.

Consider a
STEM High School

Say your middle schooler thinks she wants to be an engineer, but her public high school doesn't offer any classes in the subject. Or your 13-year-old would like to attend college, and you hope he might be able to save on the cost by earning some credit ahead of time. Across the country, there are a growing number of high schools that place a special focus on STEM, from engineering and health sciences to computer science and agriculture. While these schools can take many forms, most share a common purpose: to help students engage with the STEM fields, give them real-world skills and hands-on experiences, and expose them to rigorous, college-level work and research opportunities. "There's no one formula," but these schools strive to help teenagers understand "STEM in context," says Sharon Lynch, a professor of curriculum and pedagogy at George Washington University in Washington, D.C., who researches STEM-focused high schools.

Some schools have been around for many years, while others are brand new or redesigned programs at existing schools. Certain specialized STEM schools exist as stand-alone institutions, while others might be small magnet programs housed within a larger high school. And at a growing number of so-called dual-enrollment or early-college high schools, students can earn college credit or a full-fledged associate degree along with their high school diploma. The six-year program at the Pathways in Technology Early College High School in Brooklyn, New York, equips students with an associate degree in computer systems technology or electromechanical engineering technology.

No matter their form, most of these schools offer many opportunities to blend classroom instruction with collaborations or internships with local museums, nature centers or companies; specialized courses; intensive advising and tutoring; and other options not always available at other schools. At the Arkansas School for Mathematics, Sciences and the Arts in Hot Springs, for example, high schoolers can take classes in subjects like botany, microbiology and computer game programming. At Metro Early College High School in Columbus, Ohio, juniors and seniors studying human body systems and biomedical science might perform mock treatments or assessments on lifelike mannequins or investigate medical cases with fictional patients. At the same time, they can participate in research at Ohio State University's medical school and engage with doctors at Ohio State's Wexner Medical Center or Nationwide Children's Hospital to see some of the things they learn in the classroom firsthand. Many such schools also invite engineers, scientists and other professionals into the classroom to teach, mentor and even collaborate with students on projects.

In some cases, students at STEM schools outperform their peers at other institutions on math and science tests, and they might even be more likely to pursue college degrees in these subjects. More than 99 percent of graduates of the Illinois Mathematics and Science Academy, which opened in Aurora in 1986, have gone on to attend college, and nearly two-thirds have earned degrees in the STEM fields. But these "schools are doing way more than testing well," Lynch says. "They're really nurturing." Instead of traditional lectures, many STEM schools offer intensive lab courses or flexible schedules during the normal school day – and after school – to allow students to learn and work at their own speed. At Manor New Technology High School outside Austin, Texas, classes are project-based, encouraging collaboration among peers and often facilitated by teams of teachers (see "Learning by Doing," page 38). Students are "not just sitting there and getting information," but rather working on several cross-disciplinary projects in a given day, notes founding principal Steven Zipkes.

At certain schools, older students might help lead instruction and mentor their peers; some summer and out-of-class programs might also be available to give students some extra practice with math, say, or work on hands-on coding or robotics projects. STEM schools also still require courses in English, history, foreign language and other disciplines, sometimes embedding extra lessons in science or math within these subjects. The Arkansas School for Mathematics, Sciences and the Arts, for instance, has a class on American folk music and acoustics.

» An Arkansas School for Mathematics, Sciences and the Arts science project

tions are available. States like Texas, North Carolina and Ohio have established entire networks of STEM-focused high schools, and even some middle schools. And remember, students don't just have to attend a specialized STEM school to get a great education in those subjects. Indeed, many schools that aren't specifically STEM-focused might have cutting-edge learning opportunities and advanced courses in science, math or computer science.

Next, it's important to understand how you get in. Some schools only enroll students through a lottery system, for instance, or require an entrance exam. The Bronx High School of Science, Stuyvesant High School and several other New York City STEM schools only admit students who perform well on a comprehensive test that examines reading, math and problem-solving skills. Other schools are often referred to as "inclusive," in that they are open to all and might even make a strategic effort to recruit traditionally underrepresented populations like women, African-Americans, Latinos and low-income

To find out if a STEM-focused high school is the right fit for your child, start by searching around your area to see what institu-

students. At Manor New Tech, for example, about two-thirds of students are of African-American or Hispanic descent. At High Tech High in San Diego, where 98 percent of students go on to college, more than a third of graduates are first-generation college students.

Making a visit or speaking with someone at the school who handles admissions can be helpful. Parents might try to meet teachers and administrators and find out what courses, extracurricular opportunities and support systems – counselors and tutoring, for instance – are available to all students. One key question parents should try to understand: "What do they have within the building that will help those students along the path?" says Todd Mann, executive director of the National Consortium of Secondary STEM Schools.

But remember, just because a school has "science" or "tech" in the name doesn't mean it has all the appropriate resources to help your child succeed. A small school might have some specialized courses and internship opportunities, but it could actually offer fewer options for Advanced Placement courses than another high school. ●

Best High Schools for STEM

To determine the best STEM schools nationwide, U.S. News looked at the gold medal winners from our latest Best High Schools rankings, and then evaluated their Advanced Placement students' participation and success in AP science and math tests. Here are the top 50:

- **High Technology High School** (NJ)
- **Thomas Jefferson High School for Science and Technology** (VA)
- **BASIS Scottsdale** (AZ)
- **The Middlesex County Academy for Science, Mathematics and Engineering Technologies** (NJ)
- **Whitney High School** (CA)
- **Michael E. DeBakey High School for Health Professions** (TX)
- **The Early College at Guilford** (NC)
- **Darien High School** (CT)
- **University High School** (CA)
- **Lynbrook High School** (CA)
- **Henry M. Gunn High School** (CA)
- **Acton-Boxborough Regional High** (MA)
- **Stuyvesant High School** (NY)
- **International Academy** (MI)
- **Brookline High School** (MA)
- **Union County Magnet High School** (NJ)
- **School for the Talented and Gifted** (TX)
- **Raleigh Charter High School** (NC)
- **Palo Alto High School** (CA)
- **The Bromfield School** (MA)
- **Dover-Sherborn Regional High** (MA)
- **Academy of Allied Health and Science** (NJ)
- **International Community School** (WA)
- **Mission San Jose High School** (CA)
- **Monta Vista High School** (CA)

- **Lexington High School** (MA)
- **Blind Brook High School** (NY)
- **Saratoga High School** (CA)
- **Northside College Preparatory High School** (IL)
- **West Windsor-Plainsboro High School South** (NJ)
- **Los Gatos High School** (CA)
- **West Windsor-Plainsboro High School North** (NJ)
- **Cape Elizabeth High School** (ME)
- **Weston High School** (CT)
- **Ardsley High School** (NY)
- **Poolesville High School** (MD)
- **Westwood High School** (TX)
- **Acalanes High School** (CA)
- **Advanced Math & Science Academy Charter School** (MA)
- **Newton South High School** (MA)
- **Winchester High School** (MA)
- **D'Evelyn Junior/Senior High School** (CO)
- **Cohasset High School** (MA)
- **Staten Island Technical High School** (NY)
- **Millburn High School** (NJ)
- **Adlai E. Stevenson High School** (IL)
- **Liberal Arts and Science Academy** (TX)
- **Briarcliff High School** (NY)
- **Ithaca High School** (NY)
- **Ridgefield High School** (CT)

▶ For the rest of the rankings, visit: usnews.com/stemrankings.

Pick the Right High School Path

Why is it that fewer than 40 percent of students who enter college planning to get a STEM degree actually do? One big reason: a lack of preparation. Parents can play a big role in helping their children enjoy and excel in math and science, and not just by making sure they study and do their homework. Here's how:

First, create interest in STEM in your child

Recent research debunks the conventional wisdom that succeeding in math and science over the long haul requires special talent. "The interest drives the preparation," says Meghan Groome, executive director of education and public programs at the New York Academy of Sciences. "Anyone can do well in math." Talk with your children about how math and science are related to everyday life, suggests Judith Harackiewicz, a University of Wisconsin-Madison psychology professor. When parents do so, teens take more math and science courses in high school. Discuss, for example, how:

- Physics explains why cellphone calls are dropped in an elevator (radio waves don't travel through metal easily).
- Math leads comparison shoppers to the best cellphone rate plan.
- Chemistry helps explain how hybrid cars charge their fuel cells.

Consult a counselor as early as possible

Parents of middle school students should start looking ahead to fit in all the math and science courses colleges look for, allowing for all prerequisites. If your child isn't prepared for college math, he or she won't be able to do science or engineering.

Plan on the most demanding possible classes

You'll also want to make room for any advanced-placement courses offered that make sense given your child's strengths and interests. These challenging AP courses are designed by the College Board in different subjects to expose high schoolers to college-level material (and expectations), and this type of class is important for anyone who is planning on a STEM major.

If taking AP calculus by senior year in high school is the goal, a reasonable one for anybody who wants a STEM degree, taking algebra I in ninth grade means there will be four courses to complete in three years. Commonly, they would be geometry, algebra II, precalculus and AP calculus. Taking algebra in eighth grade opens up an additional year to fit in the advanced math. "Parents should be ensuring their kids are taking as much math as possible," says Groome, getting them after-school help or a tutor if necessary.

College admissions deans want to be able to tell from an application that a student is going to arrive on campus able to handle college-level work. "What most college admissions readers look for is a very strong precollege program, including advanced classes in math and science," says Kristin Tichenor, senior vice president at Worcester Polytechnic Institute in Massachusetts. Besides that math foundation, students who anticipate pursuing physics or chemistry, say, will want to choose the most advanced options available in those areas and in related sciences as well. And they should have a lab course or two under their belt.

What many people don't realize is that the high school guidance office sends admissions staffers a profile of the school's course offerings along with a teen's transcript. So colleges can tell if applicants have bypassed honors and AP courses in favor of easy A's. A hard-earned B in AP Biology will be more impressive than an A in an on-level biology class.

Note that while some states align high school graduation requirements with what the state's public university system wants to see on a transcript, those are minimum requirements. Students aiming to succeed in college STEM classes would want to go well beyond the minimums.

Still, the idea is to stretch, not to break. Colleges don't expect to see an entire diet of AP classes, but rather a progression of coursework that makes sense given a student's talents and interests. In AP classes a student does take, it's important to sit for the AP exam and pass.

Ask about AP prerequisites

Often, the jump into an AP class is made from an honors, not a standard, course. And if your budding scientist or engineer wants to take AP physics, he or she may need to take precalculus first or be concurrently enrolled in AP calculus. AP chemistry may first require completion of algebra II. The College Board offers recommended course preparation for AP classes at apstudent.collegeboard.org/apcourse.

Urge your child to show some initiative outside the classroom

Being prepared academically is key, but admissions officials also really like to see students who are self-starters and leaders, and who are engaged in the community outside the classroom. Ways your son or daughter can demonstrate both include:
- Taking a community college class in a subject the high school does not offer
- Enrolling in a summer academy in a STEM subject
- Choosing extracurricular activities such as Science Olympiad or Math Club
- Doing a STEM-focused internship
- Starting a STEM-related club
- Taking the SAT Subject Test in a field of interest, even if colleges don't require it. ●

Cool Contests and Competitions

>> Science, math and engineering-oriented contests, fairs and events offer unique opportunities for students to show their stuff either on their own or as a member of a team. Some competitions require entrants to represent a school or club, which might pay the entry fee. Others are free to enter. Winners may even receive cash prizes and travel to a national competition. Check the websites for more details on each contest.

- **Intel International Science and Engineering Fair:** The world's largest precollege competition, ISEF is open to winners of local and regional science fairs affiliated with the Society for Science & the Public (student.societyforscience.org).
- **Intel Science Talent Search:** High school seniors submit reports covering their original, independent research on a chosen topic to be reviewed by scientists, engineers and mathematicians (student.societyforscience.org).
- **Conrad Spirit of Innovation Challenge:** Student teams put together a pitch and presentation for potential investors, offering a product or innovation that would address a global need in one of four key areas (conradchallenge.org).
- **Team America Rocketry Challenge:** Students design, build and fly a rocket while incorporating an added challenge, such as hovering for several seconds at 800 feet with a raw egg (rocketcontest.org).
- **eCyberMission:** Sponsored by the U.S. Army, this online competition requires students to research a problem in their community and discuss their discoveries in online forums (ecybermission.com).
- **Microsoft Imagine Cup:** Participants in the cup's three primary competitions will design and build a technology using a combination of software and hardware (imaginecup.com).
- **Future City Competition:** Using computer game "SimCity," students design a city to meet a certain need, such as adequate space

for sustainable farming to feed the entire population (futurecity.org).

- **SourceAmerica Design Challenge:** Students design a technology to assist people with disabilities in the workplace (instituteforempowerment.org).
- **The DuPont Challenge:** Students research and write 700- to 1,000-word essays about using science and technology to solve a global crisis (thechallenge.dupont.com).
- **U.S. Department of Energy National Science Bowl:** Four-student teams compete with each other, answering questions on various science and math topics, such as chemistry, physics and energy (science.energy.gov).
- **Siemens Competition in Math, Science & Technology:** Students submit a report written about team or individual research relating to astrophysics, nutritional science, genetics and more (siemens-foundation.org).
- **Google Science Fair:** Students conduct an experiment on a topic, such as food science, robotics or math, and submit their results on a website they built (googlesciencefair.com).
- **U.S. National BioGENEius Challenge:** Students submit results from research conducted in response to challenges tied to health care, sustainability or the environment (biotechinstitute.org).
- **Discovery Education 3M Young Scientist Challenge:** Students create one- to two-minute videos outlining their solution to a local or global problem (youngscientistchallenge.com).
- **Science Olympiad:** At Olympiad competitions, students compete in a series of hands-on team events in areas such as genetics, earth science and mechanical engineering (soinc.org).
- **TEAMS** (Tests of Engineering Aptitude, Mathematics and Science): Student teams answer multiple-choice and essay questions relating to real-world engineering challenges (teams.tsaweb.org).
- **MATHCOUNTS Competition Series:** Students answer math questions individually and then as part of a team; qualifying students also participate in an oral Q&A session (mathcounts.org).
- **Broadcom MASTERS:** Eligible sixth- to eighth-grade students are nominated by local or regional fairs affiliated with the Society for Science & the Public for this national team science competition. Finalists research and apply STEM solutions to challenges in daily life, such as what are the strongest sewing stitches – which impacts safety devices like parachutes and seatbelts (student.societyforscience.org).
- **National Youth Cyber Defense Competition:** In early rounds, students are sent virtual images of operating systems and asked to identify cybersecurity vulnerabilities (uscyberpatriot.org).
- **American Mathematics Competitions:** Students take timed math tests with questions varying by age group; the highest-scoring students move on to invitational competitions, such as the USA Mathematics Olympiad (maa.org). ●

ROBOTS, ROBOTS

- **FIRST:** Thousands participate annually in FIRST's four student competitions for different age groups, building robots that meet required parameters like being able to stack totes and pick up litter (usfirst.org).
- **BEST:** Student teams compete against each other by building robots that must complete certain tasks within specified time limits (bestinc.org).
- **VEX:** A maker of robotics kits and a STEM-related curriculum, Vex poses challenges for students such as building a five-foot-tall robot able to move on its own and be directed by a joystick (vexrobotics.com).

sights⁺

Linda Cureton
Chief executive officer,
Muse Technologies

What sparked your interest in STEM?

I liked math and science when I was in school. I ended up majoring in math. I got a master's and a post-master's degree, and I've been working in the field ever since. There weren't a lot of people majoring in math, but the teachers encouraged me a lot to stay with it. I went to Washington, D.C., public schools, and they had a college internship that I applied for. I got into Howard University a year early because of it. I wanted to take calculus in 12th grade. I went to a performing arts school, and they didn't have calculus.

One challenge I faced was not really knowing what I could do with a math degree. A lot of girls and young men don't know. The only thing you can do, they think, is teach math. But I ended up in computer programming. NASA happened to be looking for mathematicians who knew IBM assembly language and Fortran, so they recruited me heavily. I only stayed there a year and a half, but I came back to NASA 25 years later to be the chief information officer of Goddard Space Flight Center and later the entire agency.

How can we inspire more kids?

In trying to help encourage more girls and African-American kids to enter the STEM fields, one angle that I think is overlooked is music and art. I guess I would call it the STEAM angle. I'm on the board of directors of the D.C. Youth Orchestra Program, trying to grow a strong performing arts program, using music to tap that left brain, right brain thing and get it working together.

At 8, I started playing the piano. A lot of the kids who were really good students were also taking some musical instrument. I think that there is an affinity – how good a kid is in school and playing a musical instrument, maybe because of learning the discipline of practice and studying. And they're able to keep it when they go on to higher levels of education.

What was your biggest challenge?

I didn't have many cohorts coming up through school. There were a couple of women there, but we didn't really work together. We sort of kept our heads down and studied on our own. We didn't have much of a support system. But in spite of that, I think that a woman, when you're in a field like this, you sort of learn your way. In a sense, it was good training for the real world, for the workplace, and not having a lot of women role models.

What advice do you have for underrepresented students?

Sometimes, especially in junior high school, you sort of want to do what your friends are doing. You don't want to be that different. If you're going to be successful in life, you're going to have to learn how to break away from the pack. I always say that there's a lot of success in the path least taken. Whether it means that you're going to study math or science or be a professional musician, you're going to follow a path that not many people follow. Getting comfortable with that in junior high school or in high school prepares you for success in life. ●

My Story →

Marissa Amaya
*Watson UX Designer/
Software Developer, IBM*

I've always enjoyed math, but growing up I was more interested in the performance arts – acting, singing, dancing. I even dreamed of being on Broadway. In college, I decided to use my math skills by majoring in electrical engineering. I still wanted to pursue the arts, though, and even took a six-month internship at Walt Disney World. It was fun, but I wasn't sure how stable it would be. So I searched for something else a bit more practical while still being true to myself. After taking computer science classes, I realized that fields involving computers drew on the creativity and inspiration I used in performing, but the final products could reach many more people and have a more lasting impact. When you're designing an application, the same principles of giving a good performance apply, but with a higher level of scrutiny – which creates fun challenges for developers.

At IBM, much of my job entails dreaming up futuristic applications for the Watson computer system that once won "Jeopardy!" We're exploring how it can tackle challenges like finding more effective, personalized treatments for cancer. It's the kind of stuff people once only dreamed about in science fiction. Though my title is software developer, I think of myself more as a creative person who brings that creativity to any role I tackle. To succeed, I think you should always remain true to yourself and not let other people define what you're capable of.

Dayton Rhymes
*Systems Engineer, Integrated
Defense Systems, Raytheon*

I first thought of engineering on a family visit to see my cousin. I asked what he did, and he said he was a computer systems engineer for a defense contractor. I had close friends in the military, so I thought, "This is great. This is what I want to do." In school, math was my best subject, and I always fooled around with building blocks like Legos. Both Legos and math require you to be a problem solver to put things together, and this is what engineers do.

I did well in my classes and never had problems with teachers. Those came more from some of the kids. They used to say, "You're not black. The way you speak. The way you act." Society often has an idea of how a black person should be. I just shrugged it off. I am black, but I don't have to talk in stereotypical ways. I decided that if people were going to single me out it would be for my performance. I was one of only three black students in honors classes. I worked hard and got into Stevens Institute of Technology where I majored in mechanical engineering.

I believe in seizing opportunities, so when the National Society of Black Engineers had a career fair, I went. There was a Raytheon booth. Ever since my cousin's talk, I was interested in joining the defense industry. I introduced myself and shared my résumé. That's how I got my job. Today, I help test radar surveillance systems for the military used to detect missile launches, which has given me a chance to give back.

Map a Path For

ward

Weighing the College Option

>> **Making the right moves after high school can be complicated. Today's employers are eager to hire individuals with strong skills in problem-solving, computing, engineering and other STEM abilities, and about half of all STEM jobs don't require a four-year college degree (though the diploma can often provide a salary boost). To help you and your child understand the best way to find a dream job in STEM, consider these degree and job-training options.**

CERTIFICATES AND CREDENTIALS

"A four-year degree isn't for everyone," says LeAnn Wilson, executive director of the Association for Career and Technical Education. For many students, so-called CTE programs might help them obtain specific in-demand skills and direct experience with an employer – not to mention a real job offer. Sometimes referred to as vocational training, CTE programs are often housed at community colleges and local or regional career centers. Many local companies partner with these educational institutions to give students hands-on experience in certain fields as well as to equip their current workers with new abilities. Students at the City Colleges of Chicago, for instance, can often find internships or jobs with nearby companies like CVS, JMC Steel Group and Aon. And Columbia Gorge Community College in Oregon offers a renewable energy technology certificate and associate degree program that faculty members actually developed along with area businesses like Siemens and Portland General Electric.

CTE students can typically choose from among a series of career "clusters" focused on particular industries related to STEM, such as manufacturing; health science; and agriculture, food and natural resources. Students earn certificates (or "certifications" if awarded by a company or industry group) demonstrating ultraspecific skills like welding, medical coding or bookkeeping. Professionals with certificates earn about 40 percent more on average than those with just high school diplomas, according to the Georgetown University Center on Education and the Workforce. In some fields, like IT or electronics, they could even make more than those with related associate or bachelor's degrees. Community colleges award more than 450,000 certificates each year, and some programs can be completed in one year.

To start, students should learn as much as they can about particular jobs or industries and "backwards-map from there" to see which certificate or degree is a minimum requirement, advises Amy Loyd, executive director of the Pathways to Prosperity Network for Jobs for the Future, a Boston-based nonprofit. One helpful online resource is sciencebuddies.org, which gives in-depth snapshots of dozens of STEM careers and the training they require. For example, medical lab technicians often must possess at least an associate degree or certificate in their field. Computer programmers, on the other hand, typically must hold at least a bachelor's degree.

EXPLORE JOB POTENTIAL

At the Department of Labor's mynextmove. org, you can learn about hundreds of occupations, including those forecast to have strong demand in the years ahead, such as biomedical engineers, veterinary technicians and information security analysts. Many states also have their own online job portals. High school guidance or college counselors might also know of existing partnerships with local two- or four-year colleges – and, of course, you could ask the local community colleges or technical schools directly about their programs. It can also be valuable for students to "seek out mentors beyond their immediate known circle," Loyd says, and try to understand how they prepared for their own

careers. Increasingly, some schools are offering students the chance to take online courses as they work to obtain a certificate or to "stack" credentials along the way to a job. There is no "one-size-fits-all" approach, says Wilson, so careful research and planning are important.

ASSOCIATE DEGREE

The nation's 1,100-plus community colleges enroll more than 7 million people annually, where students of all ages take advantage of the convenience and cost savings that these institutions provide. Such colleges often offer courses at night or during flexible hours, so students with jobs can keep working while they take classes. Many associate degrees can be completed in two years or less of full-time work, and credits earned can often be applied toward a four-year bachelor's degree. Students can even retake or catch up on courses where they might have struggled during high school, such as algebra, chemistry or calculus, important foundations for STEM careers.

The average cost of tuition and fees at two-year schools runs about $3,300 annually, a third of that at public four-year institutions and roughly 10 times less than at private colleges, according to the College Board. Room and board costs are also often cheaper, especially when students choose to live with their family instead of on campus. Plus, about 60 percent of those in community colleges receive at least some form of financial aid from the federal government or their state or school. A growing number of specialized high schools even allow students to work on an associate degree at the same time that they complete a traditional high school diploma (see "Consider a STEM High School," page 47).

In addition, many community college courses involve hands-on experiences in the field. At Bluegrass Community and Technical College in Kentucky, students can complete a two-year advanced manufacturing technician degree where more than half of their learning happens on the factory floor of local Toyota or 3M plants. They might get to work with machines that build automobile parts instead of just hearing about the process. Notable people who began their careers at community colleges include movie producer and director George Lucas; Frank Cruz, co-founder of the Telemundo television network; and scientist J. Craig Venter, one of the first people to sequence the human genome.

BACHELOR'S DEGREE

Depending on one's career aspirations, a bachelor's degree from a four-year college could offer the best value in terms of job and future salary potential. According to the National Association of Colleges and Employers, the most valuable bachelor's degrees among college graduates from the class of 2015 are projected in engineering (average starting salary of $63,000), computer science ($61,300) and science and math ($56,200). Four-year programs allow students to acquire more in-depth skills or to specialize in subjects like electrical engineering or biochemistry. Those who earn STEM bachelor's degrees can expect to make half a million dollars more than those with other majors over the course of their lifetimes.

Many colleges require that students complete a certain amount of credits in math or science in high school or community college, so be sure to examine the specific requirements ahead of time. At some schools, students must also apply for programs in popular subjects, like engineering or biology. Beyond that, students can pursue a master's degree or Ph.D. if they hope to work as a college professor, say, or do intensive, high-level research. Master's programs in engineering or the sciences can be completed in one or two full-time years, while doctoral programs might take five or more. Graduate degrees often come with a bump in salary, but much can depend on the field.

If you and your child think that a four-year degree is the right fit, the stories in this chapter offer tips and advice on getting in to college and finding the right program. •

Learn On the Job

Internships and apprenticeships can be great ways to gain work experience and know-how – and try out a STEM career.

Such experiences can help students truly understand if a particular job is right for them before they commit to an academic or career path. "An internship is going to be a résumé-builder; it's going to give you credibility; it's going to give you confidence," says Lauren Berger, author of "All Work, No Pay: Finding an Internship, Building Your Résumé, Making Connections, and Gaining Job Experience." Websites like idealist. org and internships.com gather potential opportunities, and teachers and guidance counselors can also offer leads. It's a good idea to explore the websites of local colleges and companies, as they might offer such gigs. Positions might be paid or unpaid and full- or part-time.

An apprenticeship, a formal program that consists of paid on-the-job training and some related academic coursework, is another option. The term is often about four years, and students typically end up with a certificate or associate degree. Many STEM industries, such as manufacturing, construction, aerospace or IT, employ apprentices. There are more than 400,000 apprentices at some 150,000 companies in the U.S. Nearly 90 percent of those who complete such programs secure a job, and average starting salaries run $50,000 annually. Learn more at dol.gov/apprenticeship.

MyStory →

Anne Beaubrun
Epidemiologist, Amgen

My mother and I came to Miami from Haiti when I was 7. I struggled to learn English, and that may be why I gravitated to math. It's a universal language. I didn't excel in sports or music, so doing well in school helped me find my identity. In high school I took rigorous International Baccalaureate courses and ranked fourth in my class. I planned to go to college near home, but my counselor told me my high school would cover my application fees to additional schools, which was a big help. I was accepted to Duke University, which gave me generous financial aid. I only regret not applying for more scholarship money – there's so much out there if you do your research!

At Duke, I started study groups and tutored other students as a way to build relationships and to improve my communications skills – so important in the sciences. Junior year, a family medical crisis led me to become interested in how science, health care and health policy interact. I wanted to arm doctors with data to better assess the effectiveness of treatment options. I applied to summer research programs to get lab experience and to build a professional network. Many universities post opportunities online and will pay living expenses and a stipend.

I was hired one summer for a project in San Diego sponsored by Amgen, a global biotechnology company. Some eight years later, I am now at Amgen full time, helping evaluate the effectiveness of novel treatments for serious diseases. **"**

Jose Contreras
Associate Technical Artist, Electronic Arts

When I was a kid, I loved playing around with computer graphics programs. After taking a media design course in high school, I spent a lot of time in the school's computer lab working on 3-D animations using software called Maya. The computer animation community online is incredible. You can join forums and working professionals often set up animation and design challenges for participants. I accepted one posted by Jeremy Birn, now a technical director at Pixar Animation Studios, who has worked on movies like "Cars" and written a terrific book on 3-D techniques, "Digital Lighting & Rendering." Jeremy asked us to light and add texture to a virtual town. He often commented on my work, which encouraged me so much.

One day I was in a Houston airport with Jeremy's book when a guy noticed it and struck up a conversation. He worked at Texas A&M, just a few minutes from my high school. The university has an outstanding computer animation program. I later sent work samples to my A&M contact, and he pointed me to recruiters and teachers. That's how I ended up at A&M.

I'm a technical artist now at Electronic Arts, working as a bridge between engineers and artists developing games. You need math skills like linear algebra for tasks such as coloring 3-D models or creating software programs. If you like to tinker, work independently, and solve problems in ever-changing environments, it's a great job. **"**

A College Prep "To-Do" List

With some careful planning, a little research, and a lot of hard work, students can make sure their high school years position them to get into the college of their choice. Here's a handy checklist for your child:

FRESHMAN YEAR

Get set for a great high school career. It's important to remember that what lies ahead is more than just a four-year audition for college. Still, it helps to start thinking now about what admissions staffers will look for just three years down the road.

- **Seek advice and teacher feedback.** Ask someone you trust to help you map out your classes (see "Pick the Right High School Path," page 50). Grades are important in ninth grade, but rigor is key, too, so don't just go for easy A's. If you get a bad grade, read (or listen) to your teacher's comments and figure out how to do better.
- **Read voraciously.** Books, newspapers, magazines, blogs – choose what engages you and remember to look up unfamiliar words.
- **Get involved.** Not only will you de-velop talents and interests that will catch a college's eye, but also you'll find school is more fun when you have activities to look forward to.

SOPHOMORE YEAR

Focus on evolving as a learner. Besides studying the material, take note of what your teachers value and consider how you can learn more efficiently – and better.

- **Refine your route.** Look ahead to which 11th and 12th grade courses you might be interested in taking and plan to work in any prerequisites.
- **Challenge yourself (wisely).** Create a balanced schedule. You want to strive for the best possible grades, but overtaxing yourself is bound to be counterproductive.
- **Get some practice.** Will you take the PSAT this year? You'll get a better sense of where you stand if you know what is on the test before you take it. Also, consider whether an SAT subject test makes sense in the spring. If you're enrolled in an AP or honors course now, the timing may be good. Take a practice version.
- **Put together a résumé.** Start jotting down your hobbies, jobs and extracurricular activities. For now, it's just a way to keep track.
- **Make the most of your summer(s).** Work, volunteer, play sports or take a class. Find an activity that builds on a favorite subject or extracurricular interest.

JUNIOR YEAR

Your grades, test scores and activities junior year constitute a big chunk of what colleges consider for admissions. Do your best in class and truly prepare for the tests you take. This can also be the time to step forward as a leader. Explore pursuits that interest you, not just because they'll look good on an application, but also because they'll help you grow as a person.

▪ **Ask for help.** If you feel stuck in your studies and in need of a breakthrough, ask teachers, parents or friends for help in finding a new approach.

▪ **Speak up in class.** You will need to ask two junior-year teachers to write college recommendations. They can't know you without hearing your thoughts, so make sure you stand out by contributing in class.

▪ **Sleep.** The average 16-year-old brain needs over nine hours of sleep to function at 100 percent, and that's exactly where you want to be.

▪ **Plan your testing calendar.** Test scores matter (along with grades), so talk with your parents and guidance counselor about which exams to take and when, and how to prepare for them. First up, the PSAT. If your 10th-grade scores put you in reach of a national merit scholarship, it might be wise to spend concentrated time prepping. Then take the SAT or the ACT in winter or early spring. Don't worry if you don't get your ideal score; you can try again. The SAT subject tests are also an option for May or June in areas where you shine or in subjects you covered junior year.

▪ **Get involved.** It's great to show you've worked hard, are dedicated to an activity, play well with others – and can lead them. Start an arts discussion group that goes to museum openings, say, or be voted team captain.

▪ **Begin building your college list.** Once you have gotten your test scores, start putting together a list of target schools. Make use of new technology and apps – and your counselor – to aid your research. Explore college websites and resources like studentaid.ed.gov and usnews.com/bestcolleges. (While you're online, be sure to clean up your Facebook act. It might get a look.)

▪ **If possible, make some campus visits.** Spring break and summer vacation are ideal times to check out a few campuses. Attend college fairs and talk with the folks behind the tables. They can give you a feel for their school and some good future contacts.

▪ **Write.** Procrastination doesn't make for a good college essay. Aim to have first drafts done by Labor Day. Share them with an English teacher or counselor.

SENIOR YEAR

This will be a year of hard work and continued preparation for you. Colleges do take senior-year transcripts into consideration. They can and will rescind offers to students who slack off, so stay focused.

▪ **Finish testing and check the boxes.** If necessary, retake the SAT, ACT or subject tests. The early fall test dates will give you time to apply early. Also, make sure you're completing all graduation requirements as well as course requirements for your target colleges.

▪ **Ask for recommendations.** Early in the school year, ask two teachers if they are willing to write a letter of recommendation for you. Choose teachers with whom you have a good relationship and who will effectively communicate your academic and personal qualities. You will want people who can offer different perspectives on your performance. Be sure to update and polish your résumé, too; it will come in handy when you're filling out applications and preparing for admissions interviews.

▪ **Apply.** Fill out each application carefully and ask someone to look over your essays critically. Check that your colleges have received records and recommendations from your high school, and have your SAT or ACT scores officially sent in. A month from the date you submit your application, call the college and confirm that it is complete.

▪ **Follow the money.** Many colleges require that all of your financial aid application forms be turned in by February. But the earlier the better.

▪ **Make a choice.** Reach out to former classmates, friends or acquaintances currently at schools that accept you who can offer the inside scoop. Talk with alumni and check if an accepted-student reception is being held near you. Then confidently make your college choice official by sending in your deposit. Done! ●

By Ned Johnson, founder of PrepMatters (prepmatters.com) and co-author of "Conquering the SAT: How Parents Can Help Teens Overcome the Pressure and Succeed."

Work w/ Dad
on FAFSA

find a
calc.
tutor

SAT prep 3:30

Get Ready
for the New SAT

If you're reading this as the parent of a high school student who will graduate in 2016, your child is preparing to sit — or has already sat — for the familiar 2400-point SAT, complete with its fancy vocabulary words and mandatory essay. But the class of 2017 will begin prepping this year for a completely overhauled test. The College Board has announced major revisions to the PSAT and the SAT, saying the current SAT has "become disconnected" from the work of schools.

What's different

The changes, which include going back to the old 1600-point composite score that's based on 800-point math and "evidence-based reading and writing" sections, and making the essay optional, are intended to better reflect the material kids are learning (or should be learning) in high school, as well as to improve the SAT's reliability as an indicator of how prepared they are to tackle college work. The current test is designed more to get at innate abilities; its defenders think the change could weaken what they see as an effective tool to identify smart, capable students at academically weaker schools.

One big shift is the way vocabulary will be handled. Rather than test knowledge of obscure words out of context (like "cruciverbalist," "mellifluous" or "prestidigitation"), the focus will be on so-called high-utility words that appear in many disciplines, and they'll be used in a passage. For example, after reading a selection about population density that uses the word "intense," the test might ask which word has the closest meaning: "emotional," "concentrated," "brilliant" or "determined."

Some college officials think this move will let students from all backgrounds show what they really know, not just what they've memorized in prepping. But others remain a fan of the way the current test gets students to tap their critical-thinking skills and knowledge of Greek and Latin roots.

Focus on evidence

The new SAT will also require students to draw conclusions by taking account of evidence, to revise and edit text, to analyze data and interpret graphs, and to solve the types of math problems most commonly seen in college courses and the workplace. It's no coincidence, observers say, that the new test will more closely resemble the ACT, which has been growing ever more popular. (The format of the ACT isn't changing, but the company is making the optional essay a more analytical exercise and breaking out new scores measuring job skills and proficiency in STEM.) The redesigned SAT will last three hours, with an extra 50 minutes allotted for an optional essay that will require analyzing a passage and how the author builds an argument.

Another change is the elimination of the guessing penalty, the practice of subtracting points for wrong answers.

Prep and practice

High school juniors and seniors, too, can capitalize on one development that has already taken effect: the College Board's new partnership with the Khan Academy to provide free online test-prep materials. The idea is to start out by taking a practice SAT, then master the material by watching in-depth explanatory videos and answering plenty of practice questions. A personalized dashboard will allow students to keep track of their progress.

Admissions deans and college counselors see the partnership as a boon for the many kids who can't take advantage of costly test prep. "We want to assist students from all backgrounds – suburban, rural, inner city – so that they all have equal access to quality test-prep materials, and now we can," says Mike Drish, the director of admissions, recruitment and outreach at the University of Illinois–Urbana-Champaign.

Whichever test your child takes, devoting time to practice should increase his or her comfort level. But some experts advise against sitting for the real thing several times in an attempt to raise the score; some colleges may ask to see all results – and they certainly want to see kids engaged in more activities than exam prep. Try to keep the testing in perspective.

The rollout. Plus: free online help

The class of 2016 takes the current SAT. **The class of 2017** takes the redesigned PSAT, and the redesigned SAT beginning in the spring of 2016. Khan Academy prep materials are available for free at khanacademy.org/sat.

Consider applying to a test-optional college or university

Many fine colleges have concluded that they don't need test scores to make admissions decisions. Two that recently joined the group: Temple University and Bryn Mawr in Pennsylvania. The National Center for Fair & Open Testing (fairtest.org) maintains a database of some 815 schools that are "test-optional." This means either that applicants choose whether to submit scores or that the schools de-emphasize the tests. ●

A Word to Students:

You Can Conquer Test Anxiety

Test day is on the horizon, and the thought is making you sweat. How to get a grip? Take a deep breath and tell yourself it's normal to be anxious – in fact, some anxiety will provide incentive to prepare. The problem is that being really anxious about taking the SAT or the ACT may actually hurt your performance. The good news: You can do quite a bit to stop test anxiety before it stops you!

. . . .

TIP 1 Practice to get a feel for the test. However you feel about the ACT and SAT, you have to give them this: They are consistent. (Thus the name standardized tests.) So the test you take will be a heck of a lot like previous ones. You already may have a feel for what to expect if you sat for the PSAT last year, or the similar preview of the ACT, known as ACT Aspire. Now you want to work through lots of practice tests, on your own and under conditions that mimic as much as possible the conditions and the stresses of the real thing. Be strict about time. Do the whole test to develop stamina. Take it in a library to add the distractions of other people. As they say in sports, "Practice like you'll play, so you can play like you've practiced."

Research shows that one of the best antidotes for students who suffer from excessive test anxiety is more practice tests. In light of this research, The College Board recently launched a collaboration with Khan Academy to offer numerous practice opportunities for the SAT online. Free! Many students approach these tests as if they're school exams. They are not. Your job is to figure out (and get used to) the ways they differ.

. . . .

TIP 2 Learn your patterns, too. If you didn't do well on a practice test, ask: Why not? Did you run out of time? If so, where and why? Identify questions that were difficult for further practice. Did you make silly errors that you can avoid making next time? When you take the test the first time, notice if something about the exam situation affected your concentration. Showing up late or forgetting to bring snacks, for example, can throw you off balance. Recognizing your own patterns as well as those of the tests can help you avoid subpar performance.

TIP 3 | Keep the test in perspective.

Remember: You are not your test score. Knowing the meaning of "laconic" or "lugubrious" does not prove you are smart. Nor does memorizing the rules of logarithms or parallel structure. The ACT and SAT are tests of acquired skills, and you may have to do some work to acquire them. But a low score is not a measurement of you – it's an indicator of the material you haven't mastered yet. Understanding that distinction should help make the test less stressful.

TIP 4 | Get more sleep.

Ever notice how, when you are tired, your mom/friend/teacher/little brother is even more annoying than usual? Actually, you're the problem. Being well-rested lowers the odds that moderate stresses will be overblown – and drown out the calm thinking center of your brain.

Getting enough sleep also strengthens the brain pathways that help you retrieve information. When you are exhausted, the guy who goes to pull the answer from the filing cabinet in the back of your head waves you off, yawning, "Sorry! I am on a break. Come back later!" You may have learned the material, but your ability to access it becomes compromised, slowing you down or foiling you altogether.

TIP 5 | Focus on the process.

Obsessing about what your score will be is bound to jangle your nerves. As you practice and on the big day, try to laser in on the process of taking the test, the part

that you can control. Psychologist Mihaly Csikszentmihalyi found that two requirements of a top performance are "concentration on the task at hand and loss of self-consciousness." So focus on the job itself and not on your results to find your groove.

TIP 6 | Get the logistics sorted out ahead of time.

Know what you need to bring (photo ID, an acceptable calculator, say) and pack your "test kit" in advance so you can grab it and go. Be clear about where you need to be, and arrive early. The stress of running late can take a toll on your sense of control.

TIP 7 | Have a Plan B.

First of all, remember that you can take the ACT or SAT (or both) again – and again. Knowing you have several shots at success should help lower the threat level. And if you still don't like your score when all is said and done, there are hundreds of "test-optional" colleges that have concluded they don't need the tests to make admissions decisions. Seriously! Go to www.fairtest.org to learn more.

Finally, I like to tell students to get up on test day and locate their swagger. Listen to your favorite music. Wear the clothes that make you feel like all that. Think about the places in your life where you are at your best. Psychologists have found that when students take a few minutes before a big test to write about the strengths and values that make them who they are, they perform better. Good luck! ●

By Ned Johnson, founder of PrepMatters (prepmatters.com) and co-author of "Conquering the SAT: How Parents Can Help Teens Overcome the Pressure and Succeed."

College Majors That Rock!

▶ **Looking for a college major with excellent job growth potential? Here are a few your child might want to consider when searching for a college:**

BIOMEDICAL ENGINEERING

These folks stand at the intersection of the life sciences, engineering and medicine and work on such advances as artificial kidneys, "designer" blood clots that save wounded soldiers on the battlefield, and stem cells to build new blood vessels and repair damaged hearts. The Bureau of Labor Statistics estimates the field will see a 27 percent growth in jobs between 2012 and 2022. Thayer School of Engineering at Dartmouth College, Georgia Institute of Technology and the University of Michigan boast highly regarded programs in the field. The accrediting organization ABET provides a list of all schools with accredited programs at abet.org.

BIOMETRICS; FORENSIC SCIENCE

As more experts are needed to operate the tools being invented to prevent and investigate crimes, colleges have begun stepping up to fill the need. Biometrics teaches students how to build automated identification devices, such as facial recognition systems. Forensic science focuses on using technology to analyze evidence.

As biometric readers (fingerprint scanners, for example) replace photo IDs and passwords, the industry is expected to grow from $50 million in 2011 to $363 million by 2018, according to Albany, New York-based Transparency Market Research. Grads pursue careers as security consultants, intelligence analysts or biometric system designers with government agencies, defense contractors or nongovernmental entities like banks.

Students in forensics learn to use technology – including many of the systems and devices developed by biometrics – to analyze crime scenes. They might specialize in forensic biology (DNA and plant or insect analysis) or forensic chemistry or toxicology. About 15 U.S. bachelor's programs are now accredited by the American Academy of Forensic Sciences, including Penn State University, Loyola University Chicago and Texas A&M University.

COMPUTER GAME DESIGN

Over 60 percent of all time spent on mobile devices is devoted to playing games. So it's hardly surprising that over 200 colleges now offer majors in game design, development and programming.

Designers must learn skills such as animation, audio design, programming and production management. Increasingly, their expertise is also being employed to create simulated training environments in which firefighters learn to deal with chemical fires and explosions, for example, and emergency personnel respond to earthquakes or other natural disasters.

Schools offering computer gaming majors include the University of Southern California, University of Utah, George Mason University in Virginia, Rochester Institute of Technology in New York, and Drexel University in Philadelphia.

CYBERSECURITY

Large companies and governments are moving aggressively to protect their computer systems. Between 2014 and 2016, the Pentagon plans to add more than 4,000 experts to the current 900 at its Cyber Command, which is responsible for defending the nation's critically important computer networks, from the military's own to civilian power grids and financial systems. Specialists in cybersecurity can also find openings in health care, keeping medical records private; energy, where systems controlling water and power supplies are susceptible to attack; and at security services firms.

To help ensure workers have the right stuff, the National Security Agency is identifying centers of excellence in cyber operations. The first four: Dakota State University in South Dakota, the Naval Postgraduate School in California, Northeastern University in Boston, and the University of Tulsa in Oklahoma. Others with programs include the University of Maryland–Baltimore County and the Polytechnic Institute of New York University.

DATA SCIENCE; BUSINESS ANALYTICS

The International Data Corporation, a technology market research firm, says the global volume of computerized data is doubling every two years. Data science focuses on finding meaningful patterns in all that information, while business analytics looks for patterns that impact business, says Christopher Starr, chairman of the computer science department at the College of Charleston in South Carolina. By studying statistics, math and programming, graduates become "data explorers," he says, and help government agencies, consulting firms, scientific organizations and companies develop strategy, understand customer behavior or predict market trends.

Business analytics entails figuring out how companies can grow and improve their performance. Courses include computer software, math, statistics and communication skills.

SUSTAINABILITY

New and retooled environmental degree programs are placing fresh emphasis on practical problem-solving. "We saw that students were tired of the gloom and doom often discussed in environmental science and wanted to talk about the solution," says Michael McKinney, professor of geology and environmental studies at the University of Tennessee–Knoxville.

The solution-oriented curriculum spans law, business, science, resource management and ethics. A course in green engineering, for example, teaches how to design industrial systems that use fewer resources. Sustainability managers in all sorts of companies and organizations look for ways to make the institution more efficient and produce less waste and pollution. Recently, at least 17 schools added sustainability majors, including the University of South Dakota, Cornell University in New York, and Oregon State University–Cascades.

PETROLEUM ENGINEERING

New technologies such as horizontal drilling and fracking have opened up shale formations thought unproductive 10 years ago, and a new crop of petroleum engineers will be needed to tap these reserves; half of the current supply are expected to retire in the next decade. These engineers earn the highest starting salary of any recent college grads, more than $100,000.

Grads will find employment in three areas, says Robert Chase, chair of the department of petroleum engineering and geology at Marietta College in Ohio: as drilling engineers who supervise the effort to access oil or gas, as production engineers who design and install the equipment needed to produce it, and as reservoir engineers who analyze how much oil or gas can be recovered. Texas A&M, the University of Oklahoma, the University of Wyoming and the University of Alaska–Fairbanks all offer highly regarded programs.

ROBOTICS

Robotics will likely create millions of new jobs over the next decade – no wonder, as it's a field that touches virtually every walk of life. Anesthesia bots are assisting in surgery, oceanographers are using underwater robots to map the underside of Arctic ice, and NASA's robotic rovers are currently surveying the surface of Mars. "If you add the fun component to engineering studies, you basically have robotics," says David Barrett, who is professor of mechanical engineering and design at Franklin W. Olin College of Engineering in Massachusetts.

Robotics majors generally study mechanical, electrical and software engineering as well as modeling and entrepreneurship. Other schools with strong robotics programs include Worcester Polytechnic Institute in Massachusetts, Lawrence Technological University in Michigan, the University of California–Santa Cruz, and Carnegie Mellon. ●

How to Impress the
College Admissions Office

 Clara Perez knew she wanted to study architecture by freshman year of high school, when she spent hours building virtual houses for the characters in the computer game "The Sims." Her college search was equally focused; she worked with her family and an independent counselor to come up with a list of schools with strong architecture programs. When she decided after a campus visit that Syracuse University was the place, she made frequent contact with the recruiter for SU's architecture school. She scheduled an interview to show off her ability to speak on her feet. "I wasn't the girl with the strongest grades in high school, but I put myself out there," she says. She got in.

An academic passion, initiative and a proven interest in a school are key to getting your child's foot in the door – and they're only some of the attributes admissions officers seek. Says Mark Montgomery, an educational consultant in Denver: "For most kids, it's not that hard to get into college as long as you're doing the right thing." U.S. News talked with admissions officers, independent college counselors and high school guidance counselors to find out what the right thing is.

Get a head start. Plenty of kids enter high school figuring they've got lots of time to perfect their act. Not so. "We do start really paying attention to students' grades in the ninth grade year," says Rick Clark, director of admission at Georgia Institute of Technology. Beyond attending to your son or daughter's grades, your child needs to be on track to fit in all the courses needed for admission, advises Thyra Briggs, vice

Consider the curriculum.

Grades in college prep courses still carry the most weight in admissions decisions. People who can show they've successfully challenged themselves in high school are "better prepared to handle college work," says Paul Marthers, vice president for enrollment at Rensselaer Polytechnic Institute in Troy, New York. At the same time, no one expects students to take every AP class offered. It's better to excel in the highest level classes that make sense given your child's interests and aptitudes, while getting eight to nine hours of sleep a night, advises Katy Murphy, director of college counseling at Bellarmine College Preparatory in San Jose, California. "We see too many kids taking advanced courses across the curriculum who crash and burn because they've taken on too much."

To repeat: It's most important to take AP classes in areas of strength. And because the guidance counselor will send colleges a detailed profile of the high school and its curriculum, the people vetting applications will know if your son or daughter has gone after easy A's. "We don't need to see the student who intends to pursue magazine journalism in AP biology, chemistry, physics and calculus," says Laura Linn, director of admission at Drake University in Des Moines, Iowa. That person is probably better off with two advanced English courses; someone planning on majoring in engineering, on the other hand, can prioritize AP math and science.

president for admission and financial aid at Harvey Mudd College in Claremont, California. That might mean making sure they have taken advanced algebra by the end of sophomore year if a college wants to see a year of calculus, or Spanish as a freshman if schools strongly prefer four years of a foreign language.

Hone extracurricular activities.

Has your child been joining a lot of clubs in order to look impressive? Stop. "Colleges are looking for two or three things done in depth," says Murphy. The goal is to select a well-rounded class of freshmen who are passionate about their particular interests, and admissions wants to get a sense from an application who the student really is.

Even if your child needs to focus on making money during the summer, those months can often be used strategically. That doesn't imply an expensive enrichment or study abroad program. Those are no more impressive than a summer job or volunteer activity close to home, and time spent sharpening skills in a sport or artistic pursuit. The key is to follow up on academic or extracurricular interests. But best not to repeat the same experience over and over again. A dedicated baseball player who spends much of one summer practicing and playing, for example, might the next summer volunteer to help coach Little League.

Develop a smart short list.

Picking colleges requires a long look inward as well as study of all those websites. Then your child needs to consider how his or her learning style and other preferences (large lectures? discussion-intensive seminars? a tight college community? Big Ten sports?) fit with each college's strengths. Ted Spencer, director of undergraduate admissions at the University of Michigan in Ann Arbor, says your child should be able to come up with five reasons for applying to every school on the list.

Then run it by a guidance counselor to be sure your child is being realistic about the chances for admission. Some high schools have software that can tell how a student stacks up against past applicants. Make sure to include a few safe bets – and that they are places where your child will be happy. Just in case.

Consider early options.

Many colleges report an increase in the number of applicants accepted through binding early decision (meaning the students promised to attend if accepted), and even more see a jump in nonbinding early-action acceptances (students get word early, without an obligation). Montgomery always advises students to apply early decision if they absolutely know where they want to go and won't need to weigh financial aid offers, since colleges like to admit students who are a sure thing. But early decision has its disadvantages. Students have to know their first choice very early senior year. And they have less bargaining power when it comes to financial aid, says Teresa Lloyd, founder of Grosse Pointe College Admissions Counseling in Grosse Pointe Park, Michigan.

Ace the essay.

The goal in choosing an essay topic is to give a sense of personality to go with the grades and test scores – not to deliver a Big Think piece on world affairs, a common mistake. Often the best inspiration is a routine experience that can be mined for a larger theme saying something about how you tick. That's the approach taken by Max Farbman, an avid percussionist from Arlington Heights, Illinois, admitted to Yale. He sent off an essay about how he's happiest when he's

Show sincere interest. As it becomes easier for students to apply to multiple schools electronically and by using the Common Application, admissions officers are alert for "stealth candidates" who do nothing but fill out the forms. Visiting, the best way to get a feel for schools, is also the best way to show interest. If you can't visit, take advantage of local college fairs and every other option for contact.

"I made sure I stood out," says Abigail Fleming of Evansville, Indiana, who applied to 10 small liberal arts colleges and, after visiting a few, zeroed in on Beloit College in Wisconsin. She stayed in touch with the admissions office when she had questions, and even drove three hours to a college fair to connect with the school again. Her clear preference for Beloit did the trick.

part of the bigger picture of the orchestra, enjoying the moment as he waits for his parts. The essay (or an optional personal statement) also provides a great opportunity to address apparent flaws in an application – poor grades during that long bout with mono sophomore year, for example. Your child can explain how he or she bounced back.

Pay attention to details.

Optional essays? Write them. A chance to elaborate on extracurricular activities? Take it. For students on the bubble, that extra effort can make the difference in whether they are admitted. Put serious thought into the teachers your child asks for recommendations. Don't always choose one who gave your child straight A's, says Briggs. A recommendation from a teacher who watched a student struggle can show how he or she responds to challenge. Finally, keep careful track of deadlines with your child. And meet them. ●

We
Did It!

How eight high school seniors made it to college

 Want to know what your child's path to college acceptance will really be like? In 2014, U.S. News visited La Jolla High School in San Diego to ask several seniors who had recently gone through the admissions process for lessons learned along the way and for their best tips for college-bound high school students.

Set in a postcard-perfect seaside community, La Jolla High is a comprehensive high school serving about 1,650 students. Because of the district's emphasis on school choice, students from an array of San Diego neighborhoods attend La Jolla. The school provides Advanced Placement courses in 21 curricular areas; 98 percent of students graduate, on average, and 70 percent go on to four-year colleges (about 22 percent go to a two-year school). White students comprise 56 percent of the student body, with Hispanics, Asians and African-Americans accounting for most of the rest.

Check out the varied routes that eight members of the class of 2014 took to get accepted to a range of colleges nationwide:

Eric Tims

▶ After applying to three dozen schools, Tims is attending Bowdoin College in Maine – and it wasn't even one of his original 35. A high school running back, Tims started with a list of 18 possibilities, a combination of Division I and III schools that included places where he would want to play football (Harvard, Brown, Dartmouth and Yale) and places he wouldn't (the University of Colorado–Boulder and the University of Oregon). But the calls just kept coming in from football coaches.

Tims got into half of the colleges he applied to (including Colorado, Oregon, George Fox University in Oregon, and the University of Redlands and the University of the Pacific in California), but no Ivies. Harvard's coach referred him to Bowdoin, and "after doing a lot of research, Bowdoin was a no-doubter," he says. He could see that it was a strong school academically and liked the sense of community: "It was like a big family." In addition to suiting up for the Polar Bears, Tims plans to major in computer science.
GPA: 3.3 unweighted
SAT/ACT scores: 620 math;

500 critical reading; 530 writing
Extracurrics: Football and working in customer service at an auto detailing firm
Essay: How he grew personally and as a student and focused on schoolwork while aiming for a possible college football career.
Big decision: He had an option to do a fifth-year program at a boarding school for another shot at the Ivies but wasn't interested.
Parental aid: "My parents helped me a lot." Mom kept him organized, and Dad talked to coaches.

> ESTABLISH RELATIONSHIPS WITH FACULTY OR COACHES.
> —Eric Tims

Visits: He toured just three schools: Bowdoin, Redlands and Pomona. "Bowdoin looks classy," he says.
Do-over: After finding out that he was close to qualifying for an Ivy, he says he "would have worked harder" freshman and sophomore years.
Clincher: Some $46,000 of Bowdoin's annual $58,000 price tag is covered by grants and other financial aid.
Advice: Do research to find schools of interest, then reach out to faculty or coaches to establish relationships that can help you get in.

> I PUT MY STORY OUT THERE ABOUT BEING FROM A TOUGH NEIGHBORHOOD.
> —Josephine Camacho

Josephine Camacho

▶ Camacho grew up in "the projects of San Diego," she says, a 40-minute bus ride and a world away from La Jolla. A part of the history she shared with colleges was that she'd grown up in a community no stranger to "drugs, gangs and violence." Camacho wants to combine her love of working with older people and an interest in business. Determined to take advantage of all La Jolla had to offer, she pushed herself academically, concentrating on mastering English, Spanish and French and striving for good grades. And she resisted the temptation to transfer to a less rigorous school closer to home, as some kids who are bussed in from afar opt to do after a year or so.

For financial reasons, she applied to only California publics – eight Cal State campuses and the University of California schools at Berkeley, Santa Barbara and Santa Cruz. She got into seven and chose UC–Santa Cruz.
GPA: 3.4 unweighted
SAT scores: 450 math; 400 critical reading; 410 writing
Extracurrics: Cheerleading; hostess and cashier at two restaurants, volunteer as a companion with the elderly, youth group leader and member of her church choir
Essay: "I put my story out there" about growing up in a tough neighborhood.
Incentive: She wants to finish what her mother, a Mexican university-trained accountant, had hoped to do – move out of the barrio and "achieve the American Dream."
Regret: Not studying for the SAT. "My classmates all had tutors and paid consultants." She didn't have the money or time.
The finances: The $34,000 in total costs will be covered by her father's unused GI Bill

benefits and grants, some help from her parents, and private scholarships.
Advice: It helped to keep separate folders on her laptop for each application, plus one quick-reference folder for PINs, passwords and usernames.

Luis Galvan

▶ Galvan is a first generation college student, headed to UC–Berkeley to study biochemistry in hopes of becoming a doctor. With money tight and his mother supporting seven, Galvan was greatly aided by a program called Reality Changers, a San Diego nonprofit that provides youth from disadvantaged backgrounds with academic support, leadership training and assistance with application costs. Galvan credits the

program – which also helped arrange an internship with a UC–San Diego researcher working on fortifying proteins against viruses – for his landing at Berkeley and becoming a Gates Millennium Scholar. The Gates program will cover tuition and living expenses for his undergrad degree and for graduate degrees in certain disciplines.

Galvan got into several California state schools but was turned down by Harvard, Yale, Princeton and the University of Pennsylvania.
GPA: 3.9 unweighted
SAT/ACT scores: 630 math; 570 critical reading; 510 writing/26 composite
Extracurrics: Fight Against Cancer Club; Reality Changers ambassador in the community, volunteer on

> **WORK A LITTLE – ONE HOUR – EACH DAY ON YOUR ESSAYS.**
> *–Luis Galvan*

community landscaping projects and at a soup kitchen
Essay: He drew an analogy between his internship and his performance as a student.
Big push: He figures he logged 350 hours applying to colleges and for the Gates program. With the help of Reality Changers, he reworked his

Gates essay 20 times.
Regret: Putting the college essays off. Would have been better to "do a little – one hour – each day."
Heads-up: Found the College Scholarship Service Profile more challenging than the FAFSA.
Advice: Get involved in clubs freshman year so you can become an officer as a senior.

Emily Young

▶ An ACL injury led Young to become "fascinated" with biomedical engineering and biomechanics, she says. That interest led to an internship at a UC–San Diego biomechanics lab, where Young processed data and MRIs for research on rotator cuff problems. Young's interest, internship and an early action application helped her get into the Massachusetts Institute of Technology, where she will study mechanical engineering with a concentration in biology and will also be playing lacrosse. She also got into Northeastern in Boston, and applied to UC–Berkeley, UCLA and Georgetown but withdrew those applications once she heard from MIT. Young figures she put a solid month of work into the MIT application, including reading four years' worth of the university's admissions blogs, which offer advice and highlight accepted students.
GPA: 3.9 unweighted
ACT score: 33 composite
Extracurrics: Captain of field hockey and lacrosse teams, anti-bullying club; Girl Scouts, elementary-school lacrosse

THE FIRST SENTENCE OF YOUR ESSAY NEEDS TO BE INCREDIBLE.
–*Emily Young*

DOUBLE-CHECK AND TRIPLE-CHECK YOUR APPLICATION.
–*Trevor Menders*

FIND SCHOOLS THAT ARE A GOOD FIT INSTEAD OF FOCUSING ON REACHES.
–*Hallie Bodenstab*

who "inspire me to set high expectations for myself."

Smart start: Began considering schools at the end of her freshman year to improve her chances of being recruited for lacrosse.

Realization: "Coaches can't pull you through admissions" at schools like MIT.

Lesson: Her father, who hires fellows in his medical practice, showed her how little time it takes to screen an application (about 45 seconds).

Advice: It's better to focus on a couple of schools than to spread yourself thin by applying to 20. Also: "The first sentence of your essay needs to be incredible," and you need a strong finish.

Trevor Menders

▶ Menders got into his top choice, Columbia University, to major in dance (and perhaps in chemistry or math as well). He had great grades and SAT scores, but "did very little community service, did not attend a multiweek overpriced SAT prep course, and did not destroy my junior year with all AP courses," he says. "Instead, I pursued what I wanted to do, took the courses I wanted to take at the appropriate

coach, internships in a biomechanics lab and as a pastry chef (which she suspects got admissions "to look at my application for a second longer")

Essay: She described the world she comes from, a family of "engineers, psychologists and doctors"

level – and it all worked out."

He does have a clear passion for dance, though: He was a trainee with the City Ballet of San Diego for the season ending in May. Menders also got into Tulane University in New Orleans; Pomona College, UC–Berkeley and UC–Santa Barbara in California; and Vassar College and Skidmore College in New York, but not Harvard and Princeton. He was wait-listed at Duke and Vanderbilt.

GPA: 4.0 unweighted
SAT/ACT scores: 720 math; 730 critical reading; 720 writing
Extracurrics: Editor and writer on the school newspaper; city ballet preprofessional division and junior company, took classical voice lessons
Essay: Making good decisions after seeing friends making not-great ones
Noteworthy: He first scored 2100 on the SAT after prepping with a $2 used study guide. When he retook the test, he worked math problems over a weekend before prom and scored 2170 the morning after it. "The prom helped me to relax."
Biggest surprise: Admissions officers are willing to work with you. "Reaching out to these people is scary, but it's really helpful."
Worked for him: Doing one application at a time, rather than completing a number of them simultaneously.
Campus visits: Showing up paid off. On the tours, he ruled out applying to Johns Hopkins University and Amherst College because he didn't get the right "vibes."
Advice: "Double- and triple-

check your application." He had to amend his Columbia app after submitting it because he put links in the wrong place and one of seven essays under the wrong prompt.

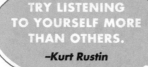

> TRY LISTENING TO YOURSELF MORE THAN OTHERS.
> –Kurt Rustin

Hallie Bodenstab

▶ Bodenstab's family hired an "expensive college counselor" to help guide the application process, leaving the budding engineer, who wants to go into sales, with mixed feelings. The counselor "kept me on track," which avoided fights with Mom and Dad, and helped her land a spot at her top choice – Lehigh University in Pennsylvania, where she is a third-generation legacy – by advising her to apply early decision.

But a downside, she feels, was that she ended up taking chemistry and AP English – a struggle – rather than regular senior English and an arts class; she loves and is active in the performing arts. She made sure she had applications ready to send to several other schools, including Lafayette College in Pennsylvania, the University of Puget Sound in Washington, and Carleton College in Minnesota, in

case Lehigh said no.

In her application, she highlighted being a strong student with an interest in science balanced by a deep résumé in theater, music and art. She also pointed to a summer computer science internship at a local firm (an experience that dissuaded her from majoring in the subject).
GPA: 3.9 unweighted
ACT score: 32 composite
Extracurrics: Drama club; singing, kickboxing, working as a singing princess at kids' birthday parties ("It pays well, and I'm never bored.")
Essay: In response to a Lehigh prompt asking her to describe what her 1 million-hit YouTube video is about: "Using my voice against authority, and doing the right thing"
Helpful resource: "The Hidden Ivies" book. "A lot of good schools don't get enough credit."
Major selling point: She was attracted to Lehigh's "real engineering" program and undergraduate research opportunities in the field.
Eye-opener: While several liberal arts schools tout their 3-2 programs – three years to get one bachelor's degree are followed by two years at a partner engineering school for a second bachelor's – it seemed too big a risk. Only the top students are likely to make the leap.
Stressor: Applying to one school while everybody else seems to be applying to 20.
Advice: Find schools that are a good fit instead of focusing your energy on reach schools.

Kurt Rustin

▶ Rustin is going to Northern Arizona University for mechanical engineering, hoping "to one day be an inventor" and hold his own patents. The track-and-field star and kayaker chose NAU for "its many labs that are about using your building skills, not just classroom smarts" because he considers himself to be more of a learn-by-doing kind of guy. "Learning on the job is definitely something I'm better at," he says.

Rustin didn't let ADHD and dyslexia hold him back, getting tutoring and extra help to succeed in AP and other challenging courses. He pursued his engineering interests by working as a hydroplane boat mechanical crew trainee and helping the Maritime Museum of San Diego build a full-scale replica of the Spanish galleon San Salvador. He also built La Jolla High's Latin department website. Rustin applied to 12 schools and got into almost all, including California Polytechnic State University, Purdue University in Indiana, Colorado School of Mines and the University of Colorado–Boulder; Bucknell University in Pennsylvania turned him down. He seriously considered CalPoly and Purdue, but had the impression they were more classroom-based than "hands-on" NAU.

GPA: 3.8 unweighted
ACT score: 31 composite
Extracurrics: Track team captain, robotics team, La Jolla High garden club; flat water and ocean kayaking
Essay: Most of his schools didn't require an essay. Responding to one prompt asking what three items he would want with him if stranded in a foreign land, he explained why he would pick a machete, the American flag and a family photo album.
Research tip: On family road

trips to areas he liked, he'd look up nearby colleges. "I'm a location guy" who thinks Flagstaff is beautiful.
Key realization: Learning shouldn't be a contest, so he didn't want to go to a school with "too much competition."
Great perk: While Rustin is an out-of-stater, he will pay tuition and fees of $14,000, about the halfway point between what in-state and out-of-state residents pay, thanks to being in one of the states in the Western Interstate Commission for Higher Education's Western Undergraduate Exchange.
Biggest surprise: "I only had to write two essays to apply to 12 colleges."
Advice: "Try to listen to yourself more than others."

Colleen Mellinger

▶ In hopes of becoming a university physics professor, Mellinger will study chemical physics at the University of California–San Diego. Although UCSD is close to home, Mellinger is still getting used to Southern California, having moved to La Jolla from Missouri before her senior year. She applied to schools where she would be able to do "a lot" of undergraduate research, beginning freshman year. "I also wanted to play

college tennis," she says.

With interests ranging from sports to science competitions and community service, she highlighted herself as well-rounded in her applications. Mellinger also got into UC–Berkeley, Case Western Reserve in Cleveland and the University of Tulsa, and was wait-listed by Washington University in St. Louis, the University of Chicago and Johns Hopkins in Baltimore. She visited several colleges and "fell in love" with UCSD.
GPA: 3.9 unweighted
SAT/ACT scores: 760 math; 600 critical reading; 700 writing/30 composite
Extracurrics: Tennis team captain, archery team captain, Science Olympiad, technology club,

academic team; volunteer with the Red Cross.
Essay: On being an inquisitive knowledge junkie: "Why? That's the question I always ask myself. Why do people do what they do? Why do magnets stick together? Why does 'knife' start with a 'k'?"
Stressor: Applying to college at the same time she was getting adjusted to a new school and new part of the country.
Off the list: Stanford. After visiting, "I didn't think it would be a good fit for me. I wanted more balance between academics and social."

> **ANYTHING YOU DO, WRITE IT DOWN, SO YOU REMEMBER LATER.**
> –Colleen Mellinger

Do-over: Would have started earlier. "It all came so fast."
Need for structure: Applying to colleges is a lesson in organization. "I'm very much of a list person."
Notable step: She took the SAT four times and the ACT three times. Overkill? Her overall scores didn't rise by a lot, but she was pleased by her improvement in various sections.
Advice: Anything you do, write it down so you can remember later to reference it on applications. She had volunteered to help clean up in Joplin, Missouri, after a tornado hit, but nearly forgot to say so. ●

MyStory →

Nino Bacani
Help Desk Technician,
Cubic Corporation

I joined the Navy after high school, became an aircraft mechanic and rose to be a petty officer. But after nine years, I wanted to spend more time with my family, so I started taking courses toward my bachelor's. I was interested in a career in information technology. With computers, new problems arise every day, so you're always learning something and building on what you already know.

But to get a job in computers with no work experience, I had to have a plan. First I found a technical school that could teach me basic computer skills and give me the certificates I needed to qualify for jobs. I researched schools online and read reviews of them. Then I chose one offering a class on troubleshooting computer hardware. Next I needed to get experience. My military transition classes gave me two great tips: Volunteer to get some on-the-job training and network, since meeting people offers you the best chance to get a job.

I made these tips part of my plan. I spent six months helping a nonprofit refurbish computers for disadvantaged families. Through a veterans group, I went to a computer conference where I chatted with a senior executive at Cubic. By coincidence, I saw a posting for a Help Desk technician at Cubic the next day. In the phone interview, I mentioned having just met the Cubic executive. I truly believe it helped me. My entire job search took about three months, and I believe that making a plan and sticking to it made the difference. **"**

Darnisha Ford
Process Control Engineer,
Georgia-Pacific

I grew up in a blended family with 19 siblings. My father passed away when I was 4, so my mother took care of us herself. I was good at math and liked to help my mother by taking apart broken appliances, troubleshooting the problem and fixing them. In high school my guidance counselor, knowing I excelled in math and science, showed me some engineering brochures. That's what I want to do, I thought, and began to think about getting an engineering degree.

I went to community college first to make sure I could do the work. I did well until I took a physics course. Physics was nothing I had ever experienced, and I had a solid F for most of the semester. The professor wasn't much help, so every day I went to the library and reread the material over and over again, praying I would get it. Finally, it started to make sense. I ended up just missing a B by two points. That class convinced me I could master anything. I ended up going to Mississippi State in Starkville to study electrical engineering.

The university has a great co-op program. A recruiter from Georgia-Pacific convinced me to come train to be a process control engineer, creating the computer programs and interfaces that allow all the machinery of the paper-making process to be operated and monitored. It was new to me, but I realized the job just required another form of learning, so I took it. Today, I write computer logic controlling GP's containerboard operation in Monticello, Mississippi. **"**

How to Get Financial Aid

Each year the federal government hands out billions in grants and loans to young people headed to college — some $138 billion in fiscal year 2013 alone. States and colleges and other organizations make big handouts, too. Here are six things families need to know to be sure of getting their share:

1. You don't have to automatically rule out high-priced colleges

New York students enjoy a relative bargain in The State University of New York and The City University of New York systems; room and board aside, tuition at the four-year colleges now runs just over $6,000 a year for state residents. Still, many colleges with large endowments want to enroll smart, capable students regardless of their ability to pay, and have the deep pockets to make that happen. A number of schools, such as Princeton, the University of Pennsylvania and the University of Chicago, have adopted no-loan or minimal-loan policies for many or all students with need. So a college with a high sticker price can sometimes be the cheapest choice.

2. You do have to fill out the FAFSA

The Free Application for Federal Student Aid determines how much a family is expected to contribute to college costs and must be completed for a student to receive any money from federal government coffers. You provide information about your income and assets (if the student is your dependent) and your child's at fafsa.gov; the result is an "expected family contribution," or EFC, that colleges use to put together an aid package. The value of your home and your retirement savings aren't held against you in the federal aid formula. Many colleges require an additional form, the CSS/Financial Aid Profile, to calculate whether your student is eligible for nonfederal awards.

3. Your child's "need" may not be what you think

The federal aid a college offers you is keyed not to how smart a student is but to need – the difference between expected family contribution and the cost of college. Families are sometimes shocked to discover that, based on income, their EFC runs to thousands of dollars that they haven't socked away. Colleges that say they "meet full need" are talking about bridging the gap between EFC and the cost – not helping out with the family contribution, too.

4. There are several parts to a financial aid package

They are: outright grants such as the Pell Grant for lower-income students (the maximum award for 2015-16 is $5,775); student loans that don't have to be repaid until college is over, the most common of which is the Stafford; and a work-study job on campus. Parents who qualify can borrow up to the full cost (minus aid received) under the PLUS loan program. State funds might be given out, too, and colleges often add "merit" aid, based not on need but on grades or leadership or musical talent, for example. Students with need get "subsidized" Stafford loans, meaning the government covers the interest until after graduation. Any student, regardless of need, can take out an unsubsidized Stafford. Currently the cap on Staffords for a typical dependent freshman is $5,500, of which $3,500 can be subsidized. (Independent freshmen can borrow $9,500.) The caps rise in later years.

5. It's critical to compare financial aid award letters carefully

Colleges have leeway in how they put their packages together and often are more generous to students they really want. A $20,000 award from one school might include $5,000 in federal and state grants, $2,500 in work-study and $12,500 in loans, while a $20,000 offer elsewhere might include $5,000 in grants, $2,500 in work-study, $2,000 in merit aid and $10,500 in loans. Borrowing less is generally better. Tip: Students may improve their chances of getting generous merit aid by applying to schools where their grades and SAT or ACT scores put them near the top of the applicant pool.

6. It's OK to ask for more

Colleges often say they don't negotiate, but many will take a second look if a family gets a better offer elsewhere. And they do want to know if your situation has suddenly changed – Mom lost her job, say, or Dad incurred big medical bills. Financial aid staffers might not be receptive to an angry phone call. But a polite request for a hearing could get results. ●

Inside "Work-Study"

If a job is part of your child's financial aid award, it can pay off in several ways. Here's what you need to know about the federal work-study program:

Amber Bunnell had always wanted to work in a library. So when the option of holding down a work-study job was offered as part of her financial aid package from Macalester College in St. Paul, Minnesota, she stated that preference "in all capital letters" on her application for a position. Working about 10 hours per week at $7.25 an hour, the Savage, Minnesota, native started as a circulation aide, manning the front desk and sorting and shelving. The next year, she was promoted to help manage the staff of about 50 undergraduate employees, with a bump in pay.

Hundreds of thousands of college students participate in the federal government's work-study program, part of its financial-aid superstructure for those who demonstrate need. The first step: Opt in when asked if your child would like to be considered for the program in question 31 of the Free Application for Federal Student Aid, or FAFSA. If your child is eligible, and if he or she follows through by actually finding a position once on campus, then your offspring can expect to earn at least federal minimum wage – currently $7.25 an hour – or the state or local baseline, if higher. Most jobs are on campus, in the dining hall, bookstore or athletic department, say, though some might be with local employers.

Some limits.
Schools are required to kick in 25 percent of every student's funding, which means that annual awards often run around $2,000 or $2,500. Students coordinate their own schedules and may work as much as they like up to the ceiling imposed by the size of the award. (An award of slightly more than $1,000 per 14-week semester would amount to 10 hours a week at $7.25 per hour, for example.)

There are "a lot of different models out there" for actually securing a job, says Joe Weglarz, executive director of student financial services at Marist College in Poughkeepsie, New York. Marist holds a work-study job fair to introduce new students to prospective employers, and the student financial services office helps facilitate the job search. At Iowa State, students have access to a jobs portal of available positions, which they apply for on their own. Other schools, like Macalester, place students in positions based on their skills and academic interests.

Besides providing some tuition or

» **Amber Bunnell's library job on campus helped her pay for school.**

spending money, work-study jobs can help students build a résumé, establish a network of mentors and potential references, and learn useful skills. Moreover, research shows that students who work about 10 to 15 hours a week tend to perform better academically, adds Desiree Noah, who coordinates student employment at La Sierra University in Riverside, California. One great perk of a work-study job is that your child's employer will probably give him or her a break when it really counts – exam time. ●

in sights+

Robert Curbeam
Raytheon engineer

As an aspiring scientist in Baltimore in the 1970s, Robert Curbeam would stand at the end of his street and marvel at NASA's Skylab space station when he could see it floating in the sky. Decades later, as an astronaut, he would see space firsthand and put his STEM skills to use installing and repairing equipment on the International Space Station. Curbeam participated in three NASA spaceflights and was the first astronaut to complete four spacewalks during a single mission. He retired from the space agency in 2007 and now serves as vice president of mission assurance for aerospace and defense company Raytheon.

What inspired you to study engineering?

When I was growing up, my mom was a chemistry teacher and I really took to it. When I started looking at colleges, I found out that I really had a keener interest in engineering. Also, when I was in middle school, I had a very good friend. He and I used to spend a lot of time together trying to design a better car or a better plane, things like that.

What kept you interested?

It was creativity mixed with mathematics. When you do design work, it's never straightforward. It almost approaches being art-like, an artistic kind of thing, where creativity and the way you think about the problem sometimes will yield a different design than someone would originally think.

How did you use that training?

There were two ways, actually. The first is the actual operation of a spacecraft. That's not to say that without a technical degree you couldn't do that. It just makes it easier to operate the spacecraft and understand the interaction between all of the systems. I also used my engineering degree after the [2003 space shuttle] Columbia accident because I was on the safety and mission assurance team that was evaluating all of the hazard analyses. At that point you really get to the nitty-gritty of how the systems work, where the hazards are, what kinds of interfaces the different systems have and their interactions, and how maybe a failure in one can cause a cascading failure in others.

What can parents do to help encourage children in STEM?

I just think it's through exposure. I count my lucky stars that I had a chemistry teacher as a mom. I had a very keen understanding of science at a very young age. I think parents need to expose their children to more of this, even if it's just going to the science museum once a year and going, "Hey, isn't this cool?" ●

A Scholarship Sampler

Scholarships to reel students into STEM are multiplying, and an online search will turn up hundreds of sources. Below is a sampling of what students with the right qualifications can compete for; some are open to all STEM students, while many target members of underrepresented groups. Requirements vary. College-bound seniors can apply for most; a few are only for students already in college.

For STEM students

- **Intertek Scholarship:** Five scholarships of up to $10,000 and an internship go to engineering students (intertek.com).
- **Buick Achievers Scholarship Program:** Fifty incoming or current college students get up to $25,000 (buickachievers.com).
- **Great Lakes Higher Education Corporation awards:** The student loan servicer offers up to 750 STEM scholarships of $2,500 (community. mygreatlakes.org).
- **The SMART Scholarship:** This Department of Defense program provides full tuition and a stipend to students in STEM willing to work for the DOD upon graduation (smart.asee.org).
- **Scholarship America Dream Award:** This renewable STEM award goes to students entering at least their second year of college (scholarshipamerica.org).
- **American Society of Civil Engineers scholarships:** Awards of $2,500 to $5,000 go to ASCE student members (asce.org).
- **American Society of Mechanical Engineers awards:** ASME offers scholarships for current undergrads and graduat-

ing high schoolers studying mechanical engineering or mechanical engineering technology (asme.org).

- **American Institute of Aeronautics and Astronautics awards:** For undergrads in aerospace-related or engineering programs (aiaa-awards.org).
- **Siemens Competition in Math, Science & Technology:** Winners and finalists get from $1,000 to $100,000 (siemenscompetition.discoveryeducation.com).
- **The Intel Science Talent Search:** Hundreds of semifinalists win $1,000 each. Finalists compete for awards up to $150,000 (student. societyforscience.org).

For women

- **Society of Women Engineers awards:** Hundreds of scholarships go to women pursuing careers in engineering, engineering technology or computer science (swe.org).
- **Google Anita Borg Memorial Scholarship:** In honor of the founder of the Institute for Women and Technology, Google awards $10,000 to students of computer science or computer engineering (google.com/edu/students).
- **HP Helion OpenStack**

Scholarship: Four scholarships of $10,000 are presented to students pursuing a career in technology (go.hpcloud.com).
- **Palantir Scholarship for Women in Engineering:** Prizes range from $1,500 to $10,000 (palantir.com/college/scholarship).

For underrepresented minorities

- **Gates Millenium Scholars Program:** Tuition and expenses, for needy undergraduate minority students in any field (gmsp.org).
- **Xerox Technical Minority Scholarship:** Up to $10,000, for minority students in a technical or engineering field (xerox.com/jobs/).
- **National Action Council for Minorities in Engineering STEM Scholarships:** For African-American, Latino or American Indian seniors in a precollege program or current college students pursuing degrees in STEM (nacme.org).
- **National Society of Black Engineers awards:** NSBE and corporate partners such as Northrop Grumman and Chevron award engineering scholarships (nsbe.org).
- **United Negro College Fund awards:** Numerous scholarship programs for students in various fields (uncf.org).
- **Society of Hispanic Professional Engineers Foundation awards:** For students of Hispanic descent (shpefoundation.org).
- **Hispanic Scholarship Fund awards:** A host of scholarships for students studying the broad spectrum of disciplines (hsf.net).
- **Great Minds in STEM scholarships:** Great Minds in STEM offers college students of Hispanic descent scholarships ranging from $500 to $10,000 (greatmindsinstem.org).
- **The Google Lime Scholarship:** Google offers $10,000 scholarships for computer science or computer engineering students with disabilities (google.com/edu/students).
- **The Generation Google Scholarship:** High school and college students in underrepresented groups are eligible for a $10,000 scholarship (google.com/edu/students).
- **The National Institutes of Health Undergraduate Scholarship:** Up to $20,000, renewable for four years, for disadvantaged students pursuing biomedical, behavioral and social science health-related research (training.nih.gov/programs).
- **Microsoft Scholarships:** Priority given to minority students, women and people with disabilities (careers.microsoft.com). ●

My Story →

Shauna Scotland
Senior Chemist, L'Oreal

In 6th grade, I had surgery on my throat to remove benign tumors, and it changed my voice. That experience made me more curious about the human body and eager to study science. My mother found an academy where I could take precollege courses like biology and chemistry, and studying for these tough classes gave me a lot of confidence. They also piqued my interest in science even more.

I learn best by doing, and in chemistry you conduct experiments to observe how different substances combine and change. In college, I gained hands-on experience through internships, including one in an environmental lab preparing and breaking down the metals in water samples for analysis. Through these internships I realized I had a passion for laboratory work. After graduation, I put my résumé on a career website and was contacted by a cosmetics company. I never fully understood how much science goes into makeup. For example, with lipstick you need the right balance of solvents, waxes and fillers for structure, and specific pigments for perfect shades. Today, I am a senior chemist in L'Oreal USA's predevelopment lab for women of color, researching new ways to formulate products for all skin types. It's always been important to me to look for new opportunities and experiences. If something interests you, then find out more about it. You can start by joining clubs, volunteering or looking for internships. Always try to push for something better, something more.

Tony Castilleja Jr.
Systems Engineer/Business Development, Boeing

My father worked long hours and my mother stayed at home, helping my siblings and me with schoolwork, urging us to work hard and get our college degrees. Since I was good at math and science in middle school, I was recommended for honors classes and special science camps, tuition-free. At one of these, I met Mae Jemison, the first African-American woman astronaut. That's when I started to think science and space were neat. A high school calculus teacher later helped me apply for NASA's SHARP apprenticeship program, where I got hands-on experience at Johnson Space Center helping with a Mars rover prototype. Listening to Mission Control during a space shuttle launch, I thought: "Wow, it would be cool to be one of those voices."

I was accepted to Rice University and did fine until I failed an engineering exam. With tutoring, though, and help from peer mentor groups (great experience for engineers, who must work well on teams!), I passed on my next try.

INROADS, a college program matching minority student interns with companies, placed me as an undergrad with Boeing helping with NASA's shuttle program. I now work on the rocket that will one day carry humans to Mars. Sadly, my mother passed away my senior year at Rice, but her support got me where I am. I remember, months after she died, sitting in Mission Control as a shuttle launched. It was the day before her birthday, and I thought: "Hey, Mom, I'm sending up a candle for you."

Finding the Right College

U.S. News publishes lists of "Best Colleges" each year to help students pick the right school. Here's a sampling from the 2015 lists and other sources to get you started:

The Biggest STEM Producers

These schools award the highest percentage of STEM degrees[†]

- California Institute of Technology
- Colorado School of Mines*
- Missouri University of Science & Technology*
- Worcester Polytechnic Institute (MA)
- Massachusetts Institute of Technology
- Rensselaer Polytechnic Institute (NY)
- Stevens Institute of Technology (NJ)
- Michigan Technological Univ.*
- Clarkson University (NY)
- Georgia Institute of Technology*
- SUNY College of Environmental Science and Forestry (NY)*
- Illinois Institute of Technology
- Carnegie Mellon University (PA)
- Stanford University (CA)
- Case Western Reserve University (OH)

[†]Based on 2013 data.

Best Engineering Programs

If you want a bachelor's degree in engineering, start with these lists

Undergrad-Focused Schools

Harvey Mudd College (CA)
Rose-Hulman Inst. of Tech. (IN)
Franklin W. Olin Col. of Engineering (MA)
United States Military Academy (NY)*
United States Air Force Academy (CO)*
United States Naval Academy (MD)*
Cal. Poly. State U.–San Luis Obispo*
Bucknell University (PA)
Cooper Union (NY)
Embry-Riddle Aeronautical U. (FL)
Villanova University (PA)

Baylor University (TX)
U.S. Coast Guard Academy (CT)*
Embry-Riddle Aeronautical U.–Prescott (AZ)
Kettering University (MI)
Lafayette College (PA)
Milwaukee School of Engineering
Santa Clara University (CA)
Swarthmore College (PA)
Union College (NY)
Univ. of Colo.–Colorado Springs*
University of San Diego

Best in the Specialties

AEROSPACE/AERONAUTICAL/ASTRONAUTICAL
Embry-Riddle Aeronautical U. (FL)
United States Air Force Academy (CO)*
Embry-Riddle Aeronautical U.–Prescott (AZ)

CHEMICAL
Rose-Hulman Inst. of Tech. (IN)
Bucknell University (PA)

CIVIL
Rose-Hulman Inst. of Tech. (IN)
United States Military Academy (NY)*
Cal. Poly. State U.–San Luis Obispo*

COMPUTER ENGINEERING
Rose-Hulman Inst. of Tech. (IN)

San Jose State University (CA)*
United States Air Force Acad. (CO)*

ELECTRICAL/ELECTRONIC/COMMUNICATIONS
Rose-Hulman Inst. of Tech. (IN)
Cal. Poly. State U.–San Luis Obispo*
United States Air Force Acad. (CO)*

INDUSTRIAL/MANUFACTURING
Cal. Poly. State U.–San Luis Obispo*

MECHANICAL
Rose-Hulman Inst. of Tech. (IN)
Cal. Poly. State U.–San Luis Obispo*
United States Military Academy (NY)*

For more information and rankings, visit: usnews.com/bestcolleges.

The Product Realization Lab at Stanford University

BRETT ZIEGLER FOR USN&WR

Research Universities

Massachusetts Inst. of Technology

Stanford University (CA)

University of California–Berkeley*

California Institute of Technology

Georgia Institute of Technology*

U. of Illinois–Urbana-Champaign*

Carnegie Mellon University (PA)

University of Michigan–Ann Arbor*

Purdue Univ.–West Lafayette (IN)*

Cornell University (NY)

Princeton University (NJ)

University of Texas–Austin*

Northwestern University (IL)

Univ. of Wisconsin–Madison*

Johns Hopkins University (MD)

Texas A&M Univ.–College Station*

Virginia Teoh*

Duke University (NC)

Pennsylvania State U.–Univ. Park*

Rice University (TX)

Univ. of California–Los Angeles*

Best in the Specialties

AEROSPACE/AERONAUTICAL/ ASTRONAUTICAL

Massachusetts Inst. of Technology

University of Michigan–Ann Arbor*

Georgia Institute of Technology*

BIOLOGICAL/AGRICULTURAL

Purdue Univ.–West Lafayette (IN)*

U. of Illinois–Urbana-Champaign*

Texas A&M Univ.–College Station*

BIOMEDICAL/BIOMEDICAL ENGINEERING

Johns Hopkins University (MD)

Duke University (NC)

Massachusetts Inst. of Technology

CHEMICAL

Massachusetts Inst. of Technology

University of California–Berkeley*

Univ. of Minnesota–Twin Cities*

CIVIL

U. of Illinois–Urbana-Champaign*

Georgia Institute of Technology*

University of California–Berkeley*

COMPUTER ENGINEERING

Massachusetts Inst. of Technology

Stanford University (CA)

Carnegie Mellon University (PA)

ELECTRICAL/ELECTRONIC/ COMMUNICATIONS

Massachusetts Inst. of Technology

Stanford University (CA)

University of California–Berkeley*

ENGINEERING SCIENCE/ ENGINEERING PHYSICS

U. of Illinois–Urbana-Champaign*

Massachusetts Inst. of Technology

Stanford University (CA)

University of California–Berkeley*

ENVIRONMENTAL/ ENVIRONMENTAL HEALTH

Stanford University (CA)

University of California–Berkeley*

Georgia Institute of Technology*

INDUSTRIAL/MANUFACTURING

Georgia Institute of Technology*

University of Michigan–Ann Arbor*

University of California–Berkeley*

MATERIALS

Massachusetts Inst. of Technology

U. of Illinois–Urbana-Champaign*

University of California–Berkeley*

MECHANICAL

Massachusetts Inst. of Technology

Stanford University (CA)

University of Michigan–Ann Arbor*

*Public schools

Best National Universities

Top-ranked schools offering a wide range of both undergrad and graduate degrees

Princeton University (NJ)

Harvard University (MA)

Yale University (CT)

Columbia University (NY)

Stanford University (CA)

University of Chicago

Massachusetts Institute of Technology

Duke University (NC)

University of Pennsylvania

California Institute of Technology

Dartmouth College (NH)

Johns Hopkins Univ. (MD)

Northwestern Univ. (IL)

Washington University in St. Louis

Cornell University (NY)

Brown University (RI)

Univ. of Notre Dame (IN)

Vanderbilt University (TN)

Rice University (TX)

University of California–Berkeley*

Best National Liberal Arts Colleges

Excellent colleges that award more than half their degrees in the arts and sciences

Williams College (MA)

Amherst College (MA)

Swarthmore College (PA)

Wellesley College (MA)

Bowdoin College (ME)

Pomona College (CA)

Middlebury College (VT)

Carleton College (MN)

Claremont McKenna College (CA)

Haverford College (PA)

Davidson College (NC)

Vassar College (NY)

United States Naval Academy (MD)*

Washington and Lee University (VA)

Colby College (ME)

Hamilton College (NY)

Harvey Mudd College (CA)

Wesleyan University (CT)

Bates College (ME)

Grinnell College (IA)

Smith College (MA)

Best Regional Universities

Top choices offering bachelor's and master's degrees and drawing heavily from their region

North

Villanova University (PA)

Providence College (RI)

College of New Jersey*

Fairfield University (CT)

Loyola University Maryland

South

Elon University (NC)

Rollins College (FL)

Samford University (AL)

The Citadel (SC)*

Belmont University (TN)

Midwest

Creighton University (NE)

Butler University (IN)

Drake University (IA)

Bradley University (IL)

Valparaiso University (IN)

Xavier University (OH)

West

Trinity University (TX)

Santa Clara University (CA)

Gonzaga University (WA)

Loyola Marymount Univ. (CA)

Seattle University

Best Regional Colleges

These schools may offer business, education and nursing as well as the liberal arts

North

U.S. Coast Guard Acad. (CT)*

Cooper Union (NY)

U.S. Merchant Marine Academy (NY)*

Elizabethtown College (PA)

Messiah College (PA)

South

Asbury University (KY)

High Point University (NC)

John Brown University (AR)

University of the Ozarks (AR)

Florida Southern College

Midwest

Taylor University (IN)

Ohio Northern University

Augustana College (SD)

College of the Ozarks (MO)

Dordt College (IA)

West

Carroll College (MT)

Texas Lutheran University

California Maritime Academy*

Oklahoma Baptist University

Master's College and Seminary (CA)

For more information and rankings, visit: usnews.com/bestcolleges.

Best Historically Black Colleges

Among the country's HBCUs, these topped the 2015 U.S. News ranking

- **Spelman College** (GA)
- **Howard University** (DC)
- **Morehouse College** (GA)
- **Hampton University** (VA)
- **Tuskegee University** (AL)
- **Xavier University of Louisiana**
- **Fisk University** (TN)
- **Florida A&M University***
- **Claflin University** (SC)
- **North Carolina A&T State Univ.***

- **North Carolina Central Univ.***
- **Tougaloo College** (MS)
- **Delaware State University***
- **Dillard University** (LA)
- **Morgan State University** (MD)*
- **Winston-Salem State Univ.** (NC)*
- **Johnson C. Smith University** (NC)
- **Clark Atlanta University**
- **Jackson State University** (MS)*
- **Elizabeth City State Univ.** (NC)*

Morehouse College in Atlanta

JIM LO SCALZO FOR USN&WR

Top Producers of Hispanic Engineers

According to Diverse: Issues In Higher Education, these schools turned out the greatest number of Hispanic engineering graduates at the baccalaureate level in 2013

- **Florida International University***
- **California State Polytechnic University–Pomona***
- **University of Texas–El Paso***
- **Texas A&M University–College Station***
- **University of Florida***
- **University of Central Florida***
- **University of Texas–Austin***
- **University of Texas–Pan American***
- **California Polytechnic State University–San Luis Obispo***
- **University of Texas–San Antonio***

- **Arizona State University–Tempe***
- **University of California–San Diego***
- **Georgia Institute of Technology***
- **University of Houston** (TX)*
- **San Diego State University** (CA)*
- **Texas Tech University***
- **University of South Florida***
- **New Jersey Institute of Technology***
- **California State University–Long Beach***
- **University of California–Irvine***

Best Values

Schools make this list based on academic quality, cost after financial aid, and how many students get aid

National Universities

School (State) (*Public)	Average cost (2013-14) after grants based on need
Harvard University (MA)	$15,169
Princeton University (NJ)	$17,994
Yale University (CT)	$17,352
Stanford University (CA)	$19,361
Massachusetts Inst. of Technology	$21,363
Columbia University (NY)	$21,906
Dartmouth College (NH)	$22,503
California Institute of Technology	$23,281
Rice University (TX)	$19,976
University of Pennsylvania	$23,542
Vanderbilt University (TN)	$21,731
Brown University (RI)	$22,162
Duke University (NC)	$23,403
Brigham Young Univ.–Provo (UT)	$12,798
Cornell University (NY)	$23,936
University of Chicago	$27,313
U. of North Carolina–Chapel Hill*	$19,614
Emory University (GA)	$23,604
Johns Hopkins University (MD)	$27,666
Northwestern University (IL)	$28,161

National Liberal Arts Colleges

School (State) (*Public)	Average cost (2013-14) after grants based on need
Amherst College (MA)	$16,590
Williams College (MA)	$18,977
Pomona College (CA)	$19,040
Wellesley College (MA)	$19,717
Soka University of America (CA)	$18,602
Vassar College (NY)	$20,211
Grinnell College (IA)	$20,653
Bowdoin College (ME)	$19,875
Washington and Lee University (VA)	$18,725
Haverford College (PA)	$20,977
Swarthmore College (PA)	$21,969
Middlebury College (VT)	$21,286
College of the Atlantic (ME)	$18,149
Colby College (ME)	$21,269
Macalester College (MN)	$24,315
Colgate University (NY)	$20,517
Claremont McKenna College (CA)	$22,911
Smith College (MA)	$24,306
Davidson College (NC)	$23,741
Hamilton College (NY)	$23,663

*Public schools

The Payback Picture

Below you'll find the schools whose students graduating in 2013 carried the lightest and heaviest debt loads, including funds borrowed from colleges, private financial institutions and federal, state and local governments. Loans taken out by parents aren't counted.

Least Debt

National Universities

School (State) (*Public)	Average amount of debt
Princeton University (NJ)	$5,558
Harvard University (MA)	$12,560
Yale University (CT)	$13,009
California Institute of Technology	$15,010
Dartmouth College (NH)	$15,660
Brigham Young Univ.–Provo (UT)	$15,769
Stanford University (CA)	$16,640
University at Buffalo–SUNY*	$17,455
University of California–Berkeley*	$17,468
U. of North Carolina–Chapel Hill*	$17,602

National Liberal Arts Colleges

School (State) (*Public)	Average amount of debt
Berea College (KY)	$6,652
Williams College (MA)	$12,474
Louisiana State Univ.–Alexandria*	$12,547
University of Virginia–Wise*	$12,772
Pomona College (CA)	$13,441
Wellesley College (MA)	$14,030
Haverford College (PA)	$14,110
Amherst College (MA)	$15,466
Thomas Aquinas College (CA)	$15,521
Vassar College (NY)	$16,365

Regional Universities

School (State) (*Public)	Average amount of debt
NORTH	
CUNY–John Jay Col. of Crim. Just.*	$11,246
CUNY–Hunter College*	$13,000
CUNY–Brooklyn College*	$14,349
SOUTH	
Hampton University (VA)	$9,878
University of North Georgia*	$12,072
Lincoln Memorial University (TN)	$12,187
MIDWEST	
Northeastern Illinois University*	$13,213
William Woods University (MO)	$15,581
Rockhurst University (MO)	$16,817
WEST	
California State U.–Fullerton*	$12,962
California State U.–Long Beach*	$13,386
California State U.–Channel Islands*	$13,791

Regional Colleges

School (State) (*Public)	Average amount of debt
NORTH	
U.S. Merchant Marine Acad. (NY)*	$5,500
Dean College (MA)	$11,678
Cooper Union (NY)	$16,640
SOUTH	
Alice Lloyd College (KY)	$8,314
North Greenville University (SC)	$15,000
Bryan College (TN)	$16,494
MIDWEST	
College of the Ozarks (MO)	$6,424
Maranatha Baptist University (WI)	$11,162
Bismarck State College (ND)*	$12,377
WEST	
Oklahoma St. U. Ins. of Tech.–Okmulgee*	$8,675
Rogers State University (OK)*	$17,153
Master's Col. and Seminary (CA)	$18,961

Most Debt

National Universities

School (State) (*Public)	Average amount of debt
Florida Institute of Technology	$41,060
Rensselaer Polytechnic Inst. (NY)	$40,584
Andrews University (MI)	$39,010
Clarkson University (NY)	$38,390
Texas Christian University	$38,317
Texas Southern University*	$37,915
Boston University	$37,694
University of Dayton (OH)	$37,551
University of St. Thomas (MN)	$36,955
St. Louis University	$36,808

National Liberal Arts Colleges

School (State) (*Public)	Average amount of debt
St. Anselm College (NH)	$42,196
Pacific Union College (CA)	$42,153
Virginia Wesleyan College	$40,804
College of St. Benedict (MN)	$40,034
Wells College (NY)	$39,965
Morehouse College (GA)	$38,136
Gordon College (MA)	$37,410
Albion College (MI)	$37,191
Wartburg College (IA)	$36,542
Siena College (NY)	$35,569

Regional Universities

School (State) (*Public)	Average amount of debt
NORTH	
St. Francis University (PA)	$50,275
Anna Maria College (MA)	$48,750
Quinnipiac University (CT)	$44,552
SOUTH	
The Citadel (SC)*	$48,862
Freed-Hardeman University (TN)	$36,434
Lynchburg College (VA)	$35,330
MIDWEST	
Rockford University (IL)	$45,577
College of St. Scholastica (MN)	$43,113
College of St. Mary (NE)	$40,026
WEST	
LeTourneau University (TX)	$44,584
Abilene Christian University (TX)	$42,585
Trinity University (TX)	$38,540

Regional Colleges

School (State) (*Public)	Average amount of debt
NORTH	
Mount Ida College (MA)	$43,860
Maine Maritime Academy*	$41,630
Delaware Valley College (PA)	$41,036
SOUTH	
Chowan University (NC)	$40,839
Tuskegee University (AL)	$39,250
Kentucky Christian University	$37,413
MIDWEST	
Adrian College (MI)	$41,763
Buena Vista University (IA)	$40,384
Mount Marty College (SD)	$38,571
WEST	
McMurry University (TX)	$39,704
Texas Lutheran University	$33,405
Menlo College (CA)	$30,134

For more information and rankings, visit: usnews.com/bestcolleges.

At Haverford, where
2013 grads averaged
$14,110 of debt

A Parent's Guide to
STEM

Executive Committee Chairman and Editor-in-Chief	Mortimer B. Zuckerman
Editor and Chief Content Officer	Brian Kelly
Executive Editor	Margaret Mannix
Managing Editor	Anne McGrath
News Editor	Elizabeth Whitehead
Art Director	Rebecca Pajak
Chief Data Strategist	Robert J. Morse
Associate Editor	Michael Morella
Director of Photography	Avijit Gupta
Photography Editor	Brett Ziegler
Contributors	Christine Cunningham, Cathie Gandel, Christopher J. Gearon, Katherine Hobson, Maura Hohman, Ned Johnson, Darcy Lewis, Margaret Loftus, Jane Porath, Ron Rohovit, Courtney Rubin
Research Manager	Myke Freeman
Prepress	Michael A. Brooks, Manager; Michael Fingerhuth

....

President and Chief Executive Officer	William Holiber
Publisher and Chief Advertising Officer	Kerry Dyer
Chief Financial Officer	Thomas H. Peck
Senior Vice President, Operations	Karen S. Chevalier
Senior Vice President, Strategic Development and General Counsel	Peter Dwoskin
Vice President, Manufacturing and Specialty Marketing	Mark W. White
Director of Specialty Marketing	Abbe Weintraub
Director of Event Sales	Peter Bowes

....

For **additional copies** of the 2015 edition of **A Parent's Guide to STEM**, contact **booksales@usnews.com**.
For all other permissions, email **permissions@usnews.com**.

NOTES

Field trips we should take

Books to get at the library

Science fair ideas

NOTES

Cool careers to think about

Colleges we should check out

Scholarships to apply for

CPSIA information can be obtained
at www.ICGtesting.com
Printed in the USA
BVHW011048091020
590699BV00008B/172

9 781931 469746